DIAGNOSTIC EXPERTISE IN ORGANIZATIONAL ENVIRONMENTS

For Dr Don Martin – M.W.

For Evan Rodney Loveday – T.L.

Diagnostic Expertise in Organizational Environments

Edited by

MARK W. WIGGINS and THOMAS LOVEDAY
Macquarie University, Australia

CRC Press
Taylor & Francis Group
Boca Raton London New York

CRC Press is an imprint of the
Taylor & Francis Group, an **Informa** business

CRC Press
Taylor & Francis Group
6000 Broken Sound Parkway NW, Suite 300
Boca Raton, FL 33487-2742

© 2015 by Mark W. Wiggins and Thomas Loveday
CRC Press is an imprint of Taylor & Francis Group, an Informa business

No claim to original U.S. Government works

Printed on acid-free paper
Version Date: 20160226

International Standard Book Number-13: 978-1-4724-3517-0 (Hardback)

Visit the Taylor & Francis Web site at
http://www.taylorandfrancis.com

and the CRC Press Web site at
http://www.crcpress.com

Contents

List of Figures

List of Contributors

Jaime Auton, Macquarie University, Australia

Peter Bruce, Swinburne University, Australia

Marino Festa, Children's Hospital at Westmead, Australia

Joseph Forgas, University of New South Wales, Australia

William S. Helton, University of Canterbury, New Zealand

Lidija Krebs-Lazendic, Macquarie University, Australia

Thomas Loveday, Macquarie University, Australia

Ben Morrison, Australian College of Applied Psychology, Australia

Natalie Morrison, Australian College of Applied Psychology, Australia

David O'Hare, University of Otago, New Zealand

Christine Owen, University of Tasmania

Nathan Perry, University of Western Sydney, Australia

Nan Xu Rattanasone, Macquarie University, Australia

Jim Rooney, University of Sydney, Australia

David Schell, Children's Hospital at Westmead, Australia

Tamera Schneider, Wright State University, United States

Ben Searle, Macquarie University, Australia

Mark W. Wiggins, Macquarie University, Australia

Preface

Where diagnosis was once intrinsically associated with medical practice, it is now being recognized as a much broader construct that has implications across a wide variety of industrial environments. The development of advanced technologies, in particular, has enabled the construction of interfaces that have brought into stark reality, the importance of information acquisition, integration, and processing, to form an assessment about the nature of a system state.

This book is intended as a comprehensive approach in which diagnosis is conceptualized as both a cognitive and social construct that largely precedes the process of decision-making. It comprises a strong theoretical component that provides the foundation for the application of diagnostic skills in medicine, finance, firefighting, air traffic control, operations centres, and criminal investigation. Importantly, it is designed to place diagnosis and diagnostic skills at the forefront of considerations in training and system design.

The theoretical basis on which the book is founded is associational learning and, more specifically, cue-based associations (Wiggins, Chapter 1) that emerge as a consequence of active, participatory experience within a specific domain. Its relationship with situational awareness is debated at both a theoretical and an applied level (O'Hare, Chapter 2), with the conclusion being drawn that cues and cue-based associations constitute the important, initial phase of situational awareness, and provide the precursor to successful comparison and projection.

To further explore the theoretical basis of diagnosis, Helton (Chapter 4) approaches the issue from a comparative psychological perspective, and uses the outcomes as a framework to consider the broader conceptual issue of diagnosis during sustained attention and vigilance tasks. The relationship between diagnosis and vigilance is an important but under-researched issue that clearly has implications for performance in high technology automated environments.

While there is an emphasis on visual diagnosis throughout the book, it is clearly the case that the cognitive principles that underscore diagnosis also need to be considered in relation to the auditory domain. This theme is considered at both the level of didactic communication (Krebs-Lazendic, Rattanasone, and Auton, Chapter 3), and in the broader context of the social endeavour (Schneider and Forgas, Chapter 6). The role of affect also emerges as part of this discussion, especially in enabling and sustaining associations in memory.

Having considered the theoretical and empirical foundation of diagnosis, interventions are considered to improve diagnostic skills from both a training and design perspective. In the context of training interventions, the emphasis is directed towards the development of strategies that facilitate the acquisition of associations that enable skilled diagnostic performance across a range of domains (Wiggins, Chapter 7). Contrasts are drawn between cue-based approaches, and other approaches to skill acquisition, including production and case-based models.

Design solutions for diagnosis are considered initially from the perspective of errors (Loveday, Chapter 5) and the nature of contemporary systems that often trigger miscues (Perry, Chapter 11), particularly in high-tempo or emergency situations. Examples are drawn from different industrial environments to demonstrate the role that designers play in ensuring that systems embody features that will cue accurate and efficient diagnosis across different contexts. Finally, solutions are offered in the form of iconic cues and diagnostic support systems that are carefully designed to contribute to skilled diagnosis.

A significant part of this book is devoted to the application of diagnostic skills in specific contexts, including major crime scene investigation (Morrison and Morrison, Chapter 9), finance (Searle and Rooney, Chapter 10), firefighting and air traffic control (Owen, Chapter 12), aviation operations control (Bruce, Chapter 13), and medicine (Schell and Festa, Chapter, 8). The authors demonstrate the application of diagnostic skills, especially in more complex environments where the responsibilities of the individual operator often yield to diagnosis at the higher-order, organizational level. Organizational culture and practice play a major role in these discussions, leading to the conclusion that diagnosis needs to be considered jointly at the level of the individual operator and at the organizational level. These discussions provide the impetus for suggestions to improve diagnostic skills, particularly in response to the development of advanced technology systems and organizational environments of increasing complexity and interdependence.

Collectively, the authors provide an overview of diagnosis that is intended to lay the foundation for future research and development, thereby enabling improvements in diagnostic skills and, ultimately, ensuring the timely and accurate response to changes in system states.

Mark W. Wiggins and Thomas Loveday

Chapter 1

Cues in Diagnostic Reasoning

Mark W. Wiggins

In most societies, expertise is revered as an exceptional ability that very few can attain and even fewer can retain over extended periods of time. No doubt, part of this reverence derives from the somewhat inexplicable nature of expert performance. Indeed, one of the more perplexing aspects of the behaviour of experts is their capacity to formulate relatively rapid, accurate responses, in situations where there is uncertainty, and often, a plethora of information. This capacity to 'cut through' superfluous detail and identify the key ingredients associated with a response represents one of society's artefacts of expertise. However, it is also a cognitive capacity that has been difficult to explain empirically.

Although expertise is generally considered a construct confined to specialists within a technical field, it is in fact, a far more ubiquitous construct since, in one way or another, humans are all experts. Whether it is making breakfast, driving a motor vehicle, or navigating computer software, expertise is relatively commonplace. However, such everyday tasks are rarely recognized as representations of expertise.

Expertise is typically regarded as an almost superhuman quality that is largely beyond the realm of the average person. For example, how is it that an expert cricket batsman can reliably make contact with a ball that is travelling 22 yards at a speed in excess of 100 kilometres per hour? Likewise, how is it possible that an expert golfer can reliably account for variations in wind direction and speed, and drive a golf ball neatly down an unfamiliar fairway?

In positioning expertise as a construct beyond the realm of the 'everyday', an opportunity has perhaps been missed to develop an understanding of the cognitive basis of the construct. For example, it might be argued that driving a ball down a fairway is a skill no more and no less complex than reliably changing from first to second gear in a manual motor vehicle or balancing consistently on a bicycle while negotiating heavy traffic. The differences in societal representations of expertise relate more to 'context' than they do to 'cognition'.

One of the difficulties in recognizing expertise as an 'everyday' construct lies in the apparently inexplicable nature of expert human performance. Much of the behaviour associated with expertise is hidden from observation so that even experts themselves have difficulty in recounting how they do the things that they do. Indeed, the difficulty for experts to verbalise the cognitive processes that underlie their performance is well-established in the empirical literature (Foley & Hart, 1990) and perhaps explains why expertise is often considered somewhat of a mystery.

The mysterious nature of expertise implies that it is also a complex phenomenon and that this complexity explains the difficulty in achieving expertise. Therefore, the expert surgeon, faced with a difficult case, is imagined to engage vast intellectual resources to integrate and thereby discern the appropriateness of one strategy over another. Moreover, this process is presumed to occur as a rational, deliberate procedure in the absence of emotion.

Despite its pervasiveness, this pedestrian view belies evidence to suggest that expertise, as a cognitive construct, is both far more common and far less complex than generally imagined. For example, there is evidence to suggest that expert cricket batsmen are not engaging in a detailed, time-consuming analysis of the anticipated trajectory of the ball, and neither are they watching the ball as it released by the bowler (Müller, Abernethy, & Farrow, 2006). There is little doubt now that batsmen are cued to the trajectory of a delivery by, amongst other indicators, the position of the fingers on the ball. Similarly, Williams, Ward, Knowles, and Smeeton (2002) have demonstrated that expert tennis players use particular features of the behaviour of an opponent to anticipate the likely to direction of the return.

The notion that experts are capable of anticipation suggests that they are employing an association between a feature that they perceive within the environment, and some form of object or event. It might be argued that, since less experienced practitioners have yet to develop these associations, this explains their inability to respond to changes in the dynamics of the interaction with the environment. Conversely, the strength of the associations amongst experts is such that they may be susceptible to deception. In a number of sports, including baseball and tennis, success is often achieved by deceiving opponents (Jackson, Warren, & Abernethy, 2006).

Deception is also used in military tactical environments, often through the use of feints, a tactic that capitalises on the expectations of opponents (Tzu & Sawyer, 1994). Historically, feints have often meant the difference between victory and defeat, even against overwhelming odds. Reputedly, Napoleon I was an accomplished proponent of the feint, often defeating overwhelmingly superior armies by feinting a frontal attack while repositioning a superior force to the flanks or the rear of the enemy (Drohan, 2006).

Such is the attraction of the feint in battle that it can be over-used to a point where, eventually, it is interpreted accurately by the enemy, and the result is a loss of any advantage that may have been gained. In the American Civil War, for example, both the Union and Confederate armies developed a propensity to anticipate the movement of enemy forces as potentially being a feint, to the extent that feints may have become less effective as the war continued (Colton, 1985). This suggests that the relationships between features and events evolve over time, and are directed towards the identification of more accurate and more efficient associations than may have previously been the case.

The Basis of Cues

There is considerable evidence to suggest that humans have a capacity to respond to stimuli in the environment in the absence of conscious deliberation (Finkbeiner & Forster, 2007). Rarely, for example, does one observe the detailed analysis of an everyday object such as a glass cup, prior to its use as a drinking vessel. The exception might involve the anthropologist who has just stumbled across shards of pottery and analyses the pieces to discern whether the vessel was used for drinking or for some other function. The difference between the two situations is context and the familiarity of the information available.

In the example of the use of a cup, the context of a kitchen might lead to the assumption that a glass cup, similar to those that might be expected in a kitchen, is indeed a drinking vessel. Establishing the context in this way tends to improve the efficient and effective retrieval of information from memory. Theoretically, this capacity is presumably a reflection of the categorization of information during information encoding. Much like a filing system, the categorization of memory systematically restricts the number of paths necessary to acquire the most appropriate task-related information (Dror, Stevenage, & Ashworth, 2008).

The notion that memories are associated with a context suggests that features associated with an environment may trigger the activation of a particular memory set which, in turn, enables the activation of specific feature–object/event relationships. This successive refinement of information retrieval is an adaptive mechanism that also enables the activation of associated memory structures that may provide the basis for expectations concerning the likelihood of events in the future.

The structure of memories remains ubiquitous, despite intensive investigation of the issue over a number of years. For some researchers, memories are assumed to embody different types of knowledge that have greater or lesser association with an event (Taatgen & Lee, 2003). For example, declarative knowledge is presumed resident in memory in the form of facts, and is typically organized around broad categories (Hayes & Heit, 2004). In contrast, procedural knowledge lies resident in the form of associations or productions that are organized around specific tasks or situations (Anderson, 1987). These productions comprise condition and action statements that, effectively, enable the association to be drawn between features of the environment and a subsequent response. In doing so, it presumably becomes possible to predict and thereby respond to changes in the environment.

Anderson (1987) suggests that the components of productions will change as new information is acquired and integrated. For example, productions may initially be constructed in the form of relatively general associations. Through repeated interactions, it may become clear that the production is more effective in some situations and less effective in others. As a result, the production becomes refined to take account of particular situations.

In refining a production, both the condition statement and the action statement may be altered to generate a more efficient production. Part of this refinement involves the identification of key features that may be evident in one production but not in others. This theoretical perspective is consistent with the notion that environmental features play a role in the efficient and effective interpretation of task-related information (Vicente & Wang, 1998).

The relatively rapid response latency associated with the activation of productions in memory may be explained through separate memory structures for declarative and procedural memory (Rehder, 2001). While declarative knowledge can be triggered rapidly, drawing inferences from declarative knowledge, in the absence of productions, means that the information must be relocated in short-term memory initially, in an attempt to search for features that may trigger an association. Since short-term memory is capacity-limited and the retention of information tends to be transient, the relocation of declarative knowledge into short-term memory results in an increase in subjective perceptions of mental workload also referred to as cognitive load (Lippa, Klein, & Shalin, 2008).

According to Anderson's (Anderson, 1993) ACT-R architecture, the development of productions represents an adaptive mechanism in which associations between features and events are acquired and retained in memory and then activated with minimal demand for working memory resources. The notion that there exists two mechanisms for the retrieval of information from memory has been interpreted by some theorists as evidence of parallel mechanisms of information processing. Such dual-process models are based on the assumption that some information is interpreted relatively slowly and consciously, while other information is interpreted rapidly and non-consciously (Gonzalez & Thomas, 2008).

Given the role of cues as a mechanism to limit the amount of information that needs to be processed during diagnosis, it might be expected that experienced diagnosticians would access relatively less information than their inexperienced counterparts during information acquisition. However, a distinction needs to be made between the initial assessment of information and the more cognitively demanding process of information acquisition. For example, it might be argued that where a list of possible sources of information is provided to the diagnostician, the initial process involves an assessment of the information sources that are available. At some point during this process, the values of the information available are acquired. This is a more resource-intensive process, since it requires more active engagement with the environment.

Evidence to support the distinction between the initial assessment of the sources of information available and the active acquisition of task-related information can be drawn from an apparent contradiction in the literature whereby some researchers have described experienced diagnosticians as acquiring extensive amounts of information (e.g. Bowling, Khasawneh, Kaewkuekool, Jiang, & Gramopadhye, 2008; Fitts, 1966), while other studies have suggested that experienced practitioners acquire relatively limited amounts of task-related information (Gegenfurtner, Lehtinen, & Säljö, 2011; Wiggins & O'Hare, 1995).

There are a number of potential explanations for these differences, including the nature of the problem being confronted, the stage of information processing under examination, and the nature of diagnosis itself.

There is a great deal of evidence available to suggest that expertise is highly domain-specific and that presenting experts with a problem outside their domain can result in behaviour that is more consistent with *proficiency* than with *expertise* (Weiss & Shanteau, 2003). Shanteau (1988) even goes so far as to suggest that the presentation of an otherwise familiar problem, in a form that is not consistent with the form in which it would occur in the operational context, presents difficulties for the expert. Similarly, expectations that involve experts having to retain information in memory or conduct calculations that are different from those that are ecologically valid will impose demands that may degrade the advantages afforded by expertise. Evidence to support these contentions can be drawn from DeGroot's famous chess experiments in which experts were able to recreate the positions of pieces on a chessboard only if the pattern of pieces was one that might emerge in the context of playing a game (de Groot, 1965). Where the pieces were presented in a random array, the performance of experts tended to degrade significantly.

At a pragmatic level, the acquisition of expertise is normally associated with a relatively greater opportunity for exposure to more frequent associations between environmental stimuli. Consequently, there is also an opportunity for the acquisition of associations between stimuli, and the possibility of greater levels of precision between associations. It is this capacity to not only 'see' an association, but to be able to distinguish different levels and types of association, that marks the expert from the novice performer (Williams, Haslam, & Weiss, 2008).

If the development of associations between stimuli represents a foundation for expertise, then the rate at which expertise is acquired and, indeed, whether expertise is achieved at all, is dependent upon the extent to which different sets of stimuli can be related. For example, a causal association between smoking and the incidence of respiratory-related cancers has been difficult to establish, primarily because of the range of variables that intercede between initial exposure and the development of the illness (Paul & Sanson-Fisher, 1996). In the case of a tennis player, the capacity to identify the relationship between the location of the opponent's racket and the subsequent trajectory of the tennis ball is dependent upon the capacity to identify those factors that precede and, therefore, might predict the trajectory of the ball (Rowe & McKenna, 2001).

The identification of the feature(s) that best predicts the trajectory of the tennis ball is, undoubtedly, important in the acquisition of expertise, but expertise also derives from the capacity to identify nuances in the location of the racket, as might occur when a player is attempting to deceive an opponent (Williams et al., 2002). The categorization of a feature at a relatively broad level (e.g. the racket is at a particular angle), rather than at a more detailed level (e.g. the racket is at a particular angle, but the opponent glanced in the opposite direction), provides the opportunity for miscueing.

Cues and Causality

One of the advantages associated with the identification and application of key features or cues is the opportunity that they provide to enable predictions about future events. In predicting events, there is an opportunity to exercise control over the environment and this desire for control is one of the most important bases of human behaviour (Bandura & Locke, 2003).

This control is evident in the assumptions of causality that occur in the absence of explicit evidence. For example, the gamblers' fallacy is a cognitive heuristic that is based on the assumption that an event is more likely to occur, the longer the duration between events (Dolan, Jones-Lee, & Loomes, 1995). This is a particularly pervasive heuristic, despite the fact that, in many cases, the events are independent. Such is the case in gambling, where gamblers assume that the longer the duration between payoffs, the more likely that a payoff will occur. This prompts the continuation of gambling activity, despite the losses incurred.

Like other heuristics, the gambler's fallacy is based on a propensity towards seeking causality between features and events. In fact, humans have been inferring causal relationships throughout history. Offerings to gods and changes in environmental conditions have been attributed both to the success or failure of crops or to the severity of a winter. Such was the power of these belief systems that whole civilizations have risen or fallen on the basis of what may have actually been random events that are interpreted as omens.

The desire to establish causal relationships reflected, undoubtedly, a desire to anticipate events, and thereby mitigate the consequences of any negative outcomes. In doing so, control could be exercised over the environment. Even in contemporary society, risk assessment and risk management strategies are mechanisms that are designed to enable the anticipation of negative (and positive) events. Both of these strategies are based on features in the form of indicators such as trend forecasts. The accurate interpretation of these features is presumably sufficient to enable predictions with greater levels of accuracy (Sheridan, 2008). In turn, these predictions enable the implementation of interventions that will alter the level of risk, and thereby improve the likelihood of a successful outcome.

In summary, the generation and utilization of cues is a process consistent with the evolutionary and psychological development of humans. Cues enable relatively rapid responses to familiar situations, facilitate predictions concerning future events, enable the differentiation of familiar from unfamiliar situations, reduce anxiety, and reduce the demands on cognitive resources. The outcome is a capacity to cope effectively with the range of stimuli that humans confront in everyday life.

Cues and the Cognitive Miser

As a means of reducing cognitive demands, the reliance on cues to direct human performance implies that humans are *cognitive misers*. This inference is drawn from information processing research in which humans are observed to invest relatively few cognitive resources in particular situations (Gray, 2000). For example, where a situation is especially familiar, there is relatively limited incentive to invest cognitive resources, since the solution is immediately available. Similarly, a difficult task for which the consequences are not significant will also attract relatively few cognitive resources (Yeo & Neal, 2004). Significant cognitive resources tend to be invested where the task is of moderate cognitive load and where the consequences of an incorrect response are significant (Omodei & Wearing 1994). In essence, the investment of cognitive resources represents a cost–benefit analysis where few resources are invested when the costs outweigh the benefits that might be gained.

A cost–benefit approach to cognition explains both the pervasiveness and the range of the cognitive heuristics that have been identified. Heuristics encapsulate a cost–benefit response to a situation, and can be exceedingly accurate or exceedingly inaccurate, depending upon the circumstances (Gonzalez, 2004; Svenson, 2008). The misapplication of heuristics is generally due to the misidentification of key features (Friedrich, 1993). The difficulty associated with the application of heuristics is the fact that they occur non-consciously. As a result, the application of the wrong heuristic and the misapplication of a heuristic can be difficult to control consciously. In practice, the outcome is an error. However, such errors reflect both the capabilities and the limitations associated with the application of heuristics.

There remains fervent debate within the literature concerning the relative value or otherwise of decision heuristics (Gigerenzer, Todd, & the ABC Research Group, 1999). However, the consensus appears to be the case that decision heuristics represent an adaptive mechanism that is both a blessing and a curse. Heuristics constitute a blessing, since they enable the retention in memory of feature–event/object relations (cues), and the rapid application of these association should the need arise (Broder & Eichler, 2006). In a technical sense, the development of a heuristic obviates the need for a conscious, considered response when next the situation is confronted. It simply involves the application of an association or, as Anderson (1993) would argue, a production. The application of a heuristic relegates the process beyond conscious processing, such that it is no longer a decision, but a response to environmental stimuli.

In the context of diagnosis, there is a recognition that the process can be either conscious or non-conscious, depending upon the nature of the task and the motivation and skills of the operator (Croskerry, 2009a; Stone & Moskowitz, 2011). The distinction between conscious and non-conscious processing in the context of diagnosis corresponds to the distinction between automatic and controlled cognitive processing described by Shiffrin and Schneider (1977). Controlled processing is conceptualized as a deliberate strategy that tends to be characteristic of less experienced operators. As operators gain experience, their cognitive processing becomes relatively spontaneous such that environmental stimuli trigger automatic associations in memory.

The distinction between automatic and controlled diagnosis has important implications for performance in the operational environment. For example, there is significant evidence to suggest that the application of heuristics, as occurs during automatic processing, can be subject to bias (Croskerry, 2002). Recognizing and responding to potential sources of bias can be difficult, since the underlying cognitive processes are beyond conscious control. As a consequence, there has been a significant effort to develop strategies that might invoke controlled processing during situations where automatic processing might otherwise be applied (Patrick, Grainger, Gregov, Halliday, Handley, James, & O'Reilly, 1999). This is especially important in environments such as medical diagnosis, where the consequences of erroneous diagnoses are clearly significant.

The distinction between controlled and automatic processing also implies two distinct information processing pathways, each of which functions in parallel, depending on the environment and the previous experience of the operator. Referred to as System 1 and System 2 processing, the former embodies characteristics of automated processing while the latter embodies characteristics of controlled processing (Evans, 2008). Despite widespread support for such mechanisms, it is not clear whether System 1 and System 2 processing represent unique information processing pathways or whether they represent different embodiments of the same information processing structures (Evans, 2008). Moreover, there is some suggestion that it is possible to 'switch' from one type of processing to another, although the precise mechanisms through which this occurs have yet to be established empirically.

From a theoretical perspective, the distinction between System 1 and System 2 processing suggests a qualitative distinction between the underlying cognitive structures that comprise the two types of processing. Where System 1 processing might engage non-conscious, cue-based strategies of encoding and response selection, System 2 processing involves a more deliberate, methodical approach in which information is acquired systematically from the environment. This information is subsequently compared to information resident in long-term memory and the actor engages one of a number of information processing heuristics to determine the most appropriate response (Evans, 2008).

A qualitative account, incorporating two distinct information-processing pathways, contrasts with more linear, cumulative approaches (Wickens, 2008). However, in terms of the utilization of cues to direct performance, the distinction between qualitative and quantitative approaches to information processing is a somewhat moot point. What remains consistent across the different models of information processing is the notion that humans can respond to environmental stimuli quickly and with little demand on cognitive processing resources.

Cues and Diagnosis

The value of cue utilization in information processing lies in its capacity to enable the identification of threats and opportunities rapidly and efficiently.

However, while cue-based associations may exist in memory, the capacity to apply this information in complex, time-constrained situations is dependent upon the integration of cue-based information. Therefore, the accuracy of diagnoses is dependent upon both the development of feature–event associations, and the extent to which the associations are integrated to develop more global cues.

Diagnosis is a cognitive process that enables the classification and, therefore, the recognition of situations or events (Mosier, Sethi, McCauley, Khoo, & Orasanu, 2007; Park & Jung, 2004). Successful diagnosis is dependent upon:

1. the retention of feature–event relations in memory;
2. the integration of feature–event relations to form meta-cues; and
3. the capacity to identify similarities between patterns of cues.

Like cues, meta-cues can be used to establish meaning. However, one might argue that they enable diagnoses with a greater level of precision and sophistication. For example, a single feature–event/object relationship may yield too broad a classification to be sufficiently predictive of the progression of a complex system. By contrast, too many feature–event/object relationships may increase the demands on information processing to a point where significant information is discarded and/or there is a failure to integrate information to derive sufficient meaning from the information available.

Arguably, there is an optimal number of feature–event/object relationships that can be applied within a particular situation given the limitations on working memory. In the case of production-based reasoning, Anderson (1993) posits the application of two mechanisms that are intended to improve the efficiency of learned associations, the first of which involves the generation of progressively finer discriminations between features as a means of improving their predictive validity. The second mechanism involves the integration of conditions and actions so that a single condition (feature) has the potential to trigger multiple events. For example, the deceleration of a motor vehicle might be associated with an approaching turn or with the movement of the vehicle to the right of the lane of traffic in preparation for the turn. Therefore, the driver of the following vehicle needs to be aware of both the deceleration of the vehicle ahead and its occupation of the right-hand side of the lane of the traffic.

Since a single feature can be associated with multiple events, either coincident or in sequence, it is important to establish those conditions under which an association is most likely to be evident. For example, an elevated heart rate may be associated with a number of underlying pathologies, and it is only in concert with other features that an accurate diagnosis can be established (Gattie & Bisantz, 2006).

Establishing relationships between features implies that features themselves can be associated, particularly in complex scenes where multiple features may be evident. In this case, the features are incorporated to become a meta-feature, much like a pattern of information. For example, in approaches to landing in the absence of instrumentation, pilots will form judgements as to the appropriateness of the angle and rate of descent, based on both the perceived shape of the runway and its location within the visual field of view (Prowse, Palmisano, & Favelle, 2008). In effect, the combination has subsumed the feature-associations that were evident in isolation. Arguably, however, the meta-cue incorporates a more precise association than would have been possible by relying on one or other of the feature–event associations. Moreover, it reduces the cognitive load necessary to integrate feature–event associations in working memory. The result is a more efficient and a more accurate judgement in a situation that requires relatively rapid responses to changes in environmental conditions.

As a construct, meta-cues are consistent with Anderson's (1993) principle of compilation in production-based reasoning whereby existing productions are integrated to form new productions of greater accuracy. Similarly, case-based reasoning is based implicitly on the principle that a number of environmental features can co-exist as part of a case or exemplar (Leake, 1999). It is the extent to which a match exists between the combined environmental characteristics and the case in memory that determines its activation.

The fact that both the strength and the nature of associations between features and events is flexible suggests that the acquisition of diagnostic skills demands not just exposure to diagnostic practice, but practice that enables the development of more refined feature–event/object associations. Consequently, it is the quality, rather than the quantity, of experience that will enable skill acquisition. While there are, undoubtedly, individual differences in the rate at which skills are acquired, fundamentally, it is learners' capacity to develop the association between features embedded in the environment and specific events or objects that determines the acquisition of skilled behaviour. Where learning environments are so complex that these relationships remain unclear or where learning environments fail to capture the nuances of associations between features and events, the rate of skill acquisition is likely to diminish.

From a cue-based perspective, the ideal skill-based training strategy is likely to be one that is capable of highlighting the relationships between features and events/objects, and ensuring that these features are not completely isolated from the operational context within which they will be applied. Further, the feature–event/object relations that are the target for training need to be carefully constructed to ensure that they are appropriate for both the learner's stage of skill development and the environment within which they are intended to be applied.

Conclusion

Experts, by their very nature, are capable of rapid, accurate responses in the face of considerable uncertainty. Clearly, the skills necessary to extract key information derive from an extensive repertoire of experiences that are well organized and highly predictive. The evidence suggests that these experiences can be represented in memory in a number of different forms, from detailed cases, to rules, to relatively simplistic associations between features and events or objects.

In complex, time-constrained situations, it is the quality of the relationship between features and events/objects that enables accurate and consistent diagnoses. The ability to extract key features from the environment, the capacity to associate features and events, and the capacity to recognise features subsequently, will determine both the rate of diagnostic skill acquisition and whether or not expertise can be achieved. This has significant implications for education, training, and assessment in advanced technology settings, since it suggests that, by understanding the key features and events associated with a situation, the exceptional can become the mundane and that expertise is by no means the province of the few.

Chapter 2

Situational Awareness and Diagnosis

David O'Hare

One of the less enjoyable tasks that must be undertaken by university teachers involves the marking of written examination scripts. In amongst the cramped and illegible handwriting and excruciating grammar and spelling lurks the occasional gem of well-written prose and lucid exposition. Unfortunately, some of these gems lose much of their sparkle on closer inspection as they reveal themselves to be answers to questions that were not actually set. Sometimes the student has misread or misinterpreted the question and set off on a different tack than that intended by the examiner. In other cases, faced with an obviously obtuse examiner who has unhelpfully set the 'wrong' question (i.e. one the student did not expect or has not prepared for), the student has simply ignored the actual question and produced the prepared answer. In either case, the answer, whatever its other merits, fails to answer the question set and inevitably results in a poor mark.

A sports official such as a football or rugby referee is required to quite literally keep up with the game and make correct and appropriate rulings in accordance with the laws of the game. Failure to know the laws of the game inevitably results in incorrect rulings that may affect the outcome and frustrate players and spectators alike. Even elite officials such as the professional referees who officiate in the English Premier football league have been known to completely 'forget' the laws of the game at times.[1] At other times, it is the nature of the event itself (e.g. 'was the player fouled or did he deliberately dive?') that is subject to debate, rather than the legitimacy of the action taken. In terms of the outcome, a correct-in-law response to a situation that has been misjudged or misinterpreted is as unhelpful as a failure to know the laws in the first place – both result in incorrect decisions.

An airline pilot has an extensive knowledge base of principles and procedures to deal with almost every situation that is likely to be experienced in the cockpit. For example, an engine failure at some point during the takeoff roll is associated with different responses depending on exactly when (in terms of the aircraft's speed) it occurs. Airline pilots are given frequent opportunities to practice their

---

1   On 17 October 2009 Sunderland beat Liverpool in an English Premier League football match with a shot that was deflected into the Liverpool goal by an inflatable beach ball that had (ironically) been thrown onto the pitch by a Liverpool supporter. Referee Mike Jones awarded a goal even though the laws of the game clearly stipulated that contact with an 'outside agent' should have led to the game being stopped and restarted with a drop-ball. The referee was subsequently demoted to a lower league. The result stood.

responses in flight simulators where there is no actual risk to people or machines. As in the previous examples, two kinds of errors are possible. One is forgetting (or not knowing in the first place) the required response. The other is misunderstanding or misjudging the situation and, consequently, enacting an inappropriate response. Interestingly, recent evidence (Casner, Geven & Williams, 2013) shows that when pilots encountered events that matched their presentation in training, then pilots had little difficulty in enacting the appropriate response. However, when the same events occurred unexpectedly in different formats, then there was much greater variability in responding.

All these examples illustrate the difference between simply retrieving a programmed or pre-prepared response to an unambiguous situation and the much more complex and messier process of understanding, interpreting or 'reading' a situation to determine what the appropriate response might be. This latter process has been described in different ways, leading to largely separate bodies of literature. One approach labels the process 'diagnosis' and investigates the individual, task and environmental features that contribute or detract from accurate diagnoses. The other labels the process 'Situational Awareness' (SA) and looks at a variety of factors that support and maintain an individual's SA.

Diagnostic Reasoning

Bayes's Theorem provides a very well-known normative framework for understanding diagnostic reasoning. For example, a physician is diagnosing a patient. A test for a particular disease has been conducted on the patient and the results are now known. In the light of this evidence, the physician wishes to estimate the likelihood that the patient has the disease. Bayes's Theorem provides a means of calculating this likelihood from the known information – the test result, the prevalence of the disease, and the accuracy of the test (i.e. the likelihood that a positive test result actually indicates the underlying condition). From these knowns, the remaining unknown (the likelihood that the condition is present given the test result) can be calculated using some simple algebra. In fact, according to Hastie and Dawes (2001), the ratio of the two conditional probabilities (i.e. the likelihood of the disease given the test result and the likelihood of a positive test result given the disease) is equal to the ratio of the two base rates (i.e. the disease prevalence and the prevalence of positive test results). This simple fact can be used as a quick check on the validity of the initial intuitions.

Bayes's Theorem provides an instructive guide to avoiding certain errors in reasoning where the two conditional probabilities are treated as similar or identical when they are, in fact, often completely different. Known as 'confusion of the inverse', a classic example is using the fact that almost all hard drug users also use 'soft' drugs (e.g. marijuana) to mistakenly infer that almost all soft drug users will end up using hard drugs. A simple comparison of the base rates (hard drug use is rare; soft drug use is common) shows that this reasoning is false (Hastie & Dawes, 2001).

While Bayes's Theorem is helpful in explicating many errors in diagnostic reasoning, it is not easy to apply nor very useful as a practical tool. It is also not very helpful in understanding the actual process of diagnostic reasoning as people most certainly do not explicitly use Bayesian reasoning. Neither do people implicitly follow Bayesian logic. For example, people generally fail to factor in base rates, or the underlying prevalence of an entity into their decisions. A classic example is provided by the 'cabs' problem (Tversky & Kahneman, 1982) where a witness to a traffic accident has identified the cab involved as blue. Knowing that eyewitness testimony is fallible, it has been determined empirically that the witness in this case correctly identifies the colour of a cab 80 per cent of the time. It is also an empirical fact that 85 per cent of the cabs in the city are green and only 15 per cent are blue. When given this information and asked to estimate the likelihood that the cab involved is actually blue, respondents typically fail to take account of the base rate prevalence of blue cabs. It can be shown easily that, despite the demonstrated eyewitness accuracy, it is actually more likely that the cab is green (Stanovich, 2010).

In fact, as David Eddy showed (Eddy, 1982, 1988), even experienced physicians regularly commit major errors in diagnostic reasoning due to a failure to follow the simple principles of Bayesian logic. Overlooking base rates or prevalence (as in the cab problem) is a common problem in medical diagnoses (e.g. Christenssen-Szalanski & Bushyhead, 1981). It can also affect judgements of sporting performance (Plessner & Haar, 2006). It can be argued that people have naturally evolved capacities for making adaptive and generally accurate decisions. As a recent invention in human history, the language of probability and probabilistic reasoning is, therefore, not something that we can deal with intuitively. Rephrasing probabilistic problems in terms of frequencies (e.g. numbers of people with a disease or symptom, etc.) can yield dramatic improvements in reasoning (Gigerenzer, 2002).

Most of the research on diagnostic problem solving in real-world domains can be described in terms of a generic information processing model (Wickens, Hollands, Banbury & Parasuraman, 2013). In the tradition of sequential stage models, the problem solver initially selectively attends to environmental cues. The selected cues thereby form the basis of: 'an understanding, awareness, or assessment of "the situation" ... a process that is sometimes labelled diagnosis' (Wickens et al., 2013, p. 248). The quality of the situational assessment or diagnosis depends on both the 'bottom-up' processes involved in cue selection and 'top-down' processes involving information previously stored in long-term memory.

Wickens et al. (2013) provide a diagrammatic representation of the 'bottom-up' processes as shown in Figure 2.1. The relationship between cues and the actual state of the world is not considered in this model, but this is covered extensively by Brunswik's (1956) 'lens model' (Figure 2.3). The model outlined by Wickens et al. addresses the issues of cue salience, reliability and diagnosticity.

Hypothesis belief

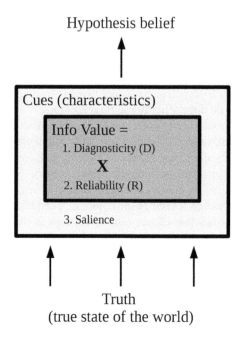

Figure 2.1 **General outline of the diagnostic process (adapted from Wickens et al., 2013)**

Salience is a rather imprecise concept generally referring to physical features that distinguish a cue – sometimes referring to the noticeability or the conspicuousness of a cue. Many properties of a cue can contribute to its salience, from its inherent psychophysical properties such as brightness or loudness, to more complex properties involving a comparison with other cues encountered previously. These properties include novelty, ambiguity and surprisingness. Berlyne (1971) labelled these 'collative' as they involve collating or comparing information from different sources.

Cue reliability refers to the credibility or believability of a cue. For example, information emanating from a consultant physician tends to carry greater credibility than information emanating from a candidate in a political campaign. Diagnosticity refers to the value of a cue in discriminating between alternative hypotheses. A cue that is associated with multiple hypotheses is not very diagnostic, whereas a cue that is associated with only one hypothesis is highly diagnostic. For example, in medical diagnosis, the presence of a slight fever is normally not very diagnostic since many conditions present with slight fevers. In a process known as 'differential diagnosis' a physician may seek out further tests that are more diagnostic (i.e. associated with fewer conditions) so that possible conditions can be eliminated and the only remaining hypothesis is likely to be the correct one.

Wickens et al. (2013) summarise a number of 'vulnerabilities' or sources of bias that may arise from these 'bottom-up' processes in the diagnostic task. Chief among these are overweighting or overprocessing salient cues and the non-optimal weighting of the value of cues – in effect, ignoring a cue's reliability or diagnosticity. Meehl (1954) provided an early demonstration of the failure of unaided clinical judgement to predict outcomes as well as simple linear models combining a small number of diagnostic cues. Meehl reviewed 20 studies in which a linear model and human judges predicted a global outcome such as academic grades, imminent dangerousness, or the outcome of medical treatment. Not one study found that human judges outperformed the statistical model. Over 40 years later, Meehl published another review (Grove & Meehl, 1996) of the literature that had now grown to 136 published studies involving over 600 comparisons between unaided human judgement and a mechanical or statistical prediction. The comparisons were restricted to studies where both the judges and the model used identical information or where the judges had additional information to that included in the model. As was evident 40 years earlier, the overwhelming majority (94 per cent) of the studies reported either no difference between the two methods or more accurate predictions from the mechanical linear model. Dawes (Dawes & Corrigan, 1974; Dawes, 1979) demonstrated that even randomly weighted cues could often outperform complex multiple regression models in predictive tasks. Unit weighting (with each cue weighted +1 or -1) of cues in a linear model resulted in even greater accuracy (Dawes, 1979; Hastie & Dawes, 2001).

The well-known 'heuristics and biases' approach to judgement and decision making (Tversky & Kahneman, 1974) describes a number of 'top-down' influences from long-term memory on humans' ability to engage in diagnostic reasoning and probabilistic forecasting. The three most well-known examples are the availability, anchoring and adjustment, and representativeness heuristics (Tversky & Kahneman, 1974). Many more heuristics have been added subsequently to the list, including the recognition heuristic (Goldstein & Gigerenzer, 2002), the take-the-best heuristic (Gigerenzer & Goldstein, 1999), and the affect heuristic (Slovic, Finucane, Peters & MacGregor, 2007). In addition, the view that human reasoning is invariably fatally compromised by the use of heuristics has come under increasing challenge recently. Heuristic processes have been shown to lead to more accurate judgements than algorithmic computations in appropriate environments (Gigerenzer & Brighton, 2009).

Situational Awareness

Work on Situational Awareness (SA) has been heavily dominated by the influence of one model proposed by Endsley (1995a). The definition of SA proposed by Endsley (1995a) has been reproduced extensively and has become the canonical definition of the concept: 'The perception of the elements in the environment within a volume of time and space, the comprehension of their meaning, and the projection of their status in the near future' (p. 36). Each of the three parts (perception, comprehension and projection) was further defined as a level of SA in a hierarchical ordering where the higher level draws information from the lower level(s). In plain terms, the three levels correspond to 'What is it doing, why is it doing that, what will it do next?' (p. 38).

Endsley's (1995a) extensive elaboration of the SA construct and its enabling processes makes it very clear that SA is a 'person's state of knowledge about a dynamic environment' (p. 60) brought about by a number of information processing activities. Somewhat confusingly, SA is simultaneously shown as the first stage in a traditional stage model of information processing (Endsley, 1995a, Fig. 1) followed by decision making and action. Endsley (1995a) emphasises this point quite clearly: 'SA decision making, and performance are different stages' (p. 36).

Flach (1995) was an early critic of the stage/box approach: 'More boxes in models are not needed' (p. 154). Flach also draws attention to the inherent circular reasoning that accompanies many analyses of accidents in which 'loss of situational awareness' is cited as a causal factor: 'How does one know that SA was lost? Because the human responded inappropriately. Why did the human respond inappropriately? Because SA was lost. Is this keen insight or muddled thinking?' (p. 151).

An example of such muddled thinking appears in Endsley (1995a) which cites a study by Kuipers et al. (1989) on fighter aircraft accidents. In a majority (56 per cent) of cases, a 'lack of attention to primary flight instruments' was a factor. In over a quarter of cases (28 per cent), there was too much attention paid to target planes. Endsley (1995a, p. 42) concludes: 'Focusing on only certain elements led to a lack of SA and fatal consequences'. It is difficult to see what value this conclusion adds to the already established 'lack of attention to primary flight instruments' as the reason why these aircraft crashed. In fact, in a quarter of the crashes, there seems to have been too much SA of one element in the environment and not enough of another.

Both Endsley (1995a) and her critics agree that the core of establishing SA lies in the process of attending to critical environmental cues and, via pattern matching, linking these to response patterns. In this sense, Endsley's (1995a) SA construct is similar to Klein's (1989) recognition-primed decision model. Flach (1995) praises the SA field for its theoretical focus on cue utilization and its recognition that the critical question is the correspondence between the perceived cues and the state of the world. This addresses the problem of ecological validity (Brunswik, 1952) generally ignored in most human factors research. Flach (1995) neatly encapsulates the contribution of the field: 'Discussions of SA have prompted attention not only to what is inside the head (awareness) but also to what the head is inside of (situations)' (p. 152).

Durso and Sethumadhavan (2008) point out that the impact of the study of SA has spread far beyond its original home in the aviation domain. In particular, it has had positive impacts on our understanding of automation and on interface design (see Chapter 5).

In the past few decades, there has been intense interest in the measurement of SA. Measures are used to assess differences in design technologies, training interventions and so forth. Stanton, Salmon, Walker, Baber and Jenkins (2005) provide a very comprehensive overview of the many methods available. The best known is the one originally proposed by Endsley (1995b) known as the 'Situational Awareness Global Assessment Technique' (SAGAT). The SAGAT approach is carried out in a simulator and involves pausing the task at certain points. Information screens are blanked and the participant is probed for information relating to each of the three levels of Situational Awareness as posited by Endsley (1995a). By its nature, the SAGAT approach is invasive and dependent on the participant's ability to recall information from memory.

Capturing the state of an individual's SA through measurement techniques such as SAGAT has value in assessing, for example, the effects of different levels of automation (Endsley & Kiris, 1995). Using a simplified vehicle navigation task, Endsley and Kiris (1995) showed that intermediate levels of automation resulted in higher levels of SA and subsequent, improved manual task performance following automation failure. Measures of SA, as a product, are useful as dependent variables in studies of system design and automation (Durso & Sethumadhavan, 2008). A number of studies have provided independent evidence for the validity of SAGAT measures (e.g. Salmon, Stanton, Walker, Jenkins, Ladva, Rafferty, & Young, 2009), although concerns about the potentially invasive nature of the off-line probe technique remain (e.g. Sarter & Woods, 1995).

Other studies have focussed on SA as a process of sensemaking or situational understanding (Durso & Sethumadhavan, 2008). Consequently, studies of the processes that underlie SA are widely dispersed in various corners of the human factors literature, especially in areas relating to memory and attention (Wickens et al., 2013).

Some research has examined the individual differences that may underlie SA. Understanding why some individuals seem to have a greater capacity to attain higher levels of SA in a given domain would have important implications for selection and training. Caretta, Perry and Ree (1996) correlated peer assessments of F-15 fighter pilots' 'SA ability' with cognitive, perceptual-motor and personality measures. The best predictor of SA ability was the total number of hours flown in the F-15. When this was held constant, general cognitive ability emerged as a significant predictor of SA ability. General cognitive ability is a higher-order factor that accounts for the inter-correlations between various measures of verbal ability, spatial ability, working memory and so forth. In some sense, general cognitive ability can be equated to some kind of executive control over cognitive processes.

A purpose-built test of this executive control capacity was developed by Roscoe, Corl and LaRoche (2001) for the specific purpose of providing a valid predictor of SA ability. The WOMBAT test requires the coordinated performance of a primary dual tracking task along with a mental rotation, pattern recognition and working memory task. The tracking task can be allocated to an autopilot but this requires constant monitoring due to frequent failures. The other tasks are adaptive to the participant's abilities. To optimise their score, the participant must be constantly aware of the changing nature of individual task demands as well as the shifting payoff structure – in other words they need to maintain a high level of SA.

O'Hare (1997) found that WOMBAT scores of executive control differentiated elite soaring pilots from other, highly experienced pilots and from matched (non-pilot) controls. There is no question that success in international gliding/soaring competitions requires high levels of SA so these findings at least support the notion that there are measurable, individual cognitive abilities that underlie SA. O'Brien and O'Hare (2007) also found the WOMBAT measures of executive control to be highly predictive of scores on an air traffic control simulation.

Process-oriented studies make use of a wide variety of measures such as reaction times and eye movements to capture underlying hypothesized mechanisms. This is reflected in the associated measurement techniques such as the Situation Present Assessment Method (SPAM) developed by Durso and Dattel (2004) that uses response latency to real-time probes to assess the process of SA. A number of studies have shown satisfactory predictive validity for the on-line probe method (Bacon & Strybel, 2013).

An Alternative View of Situational Awareness

Appearing in the same special edition of *Human Factors* as Endsley's (1995a) highly cited paper was a radically different theory of Situational Awareness. Smith and Hancock (1995) proposed an adaptive ecologically-based model of goal-directed action. This instantiates Flach's (1995) observation noted above that SA cannot be understood as a solipsistic mental construct but only makes sense in the interaction between agent and environment: 'the root of SA is the adaptive capacity to guide behaviour in response to dynamic situations' (p. 140). Or, to put it another way: 'knowing what must be known and doing what must be done' (p. 144). With a citation count less than one-tenth (80 v. 861) of the Endsley paper (Web of Science, 21 November 2013), Smith and Hancock's ecologically-based view has had very much less influence on the heavily cognitive-behavioural discipline of human factors or engineering psychology. Another reason for the disparity in influence lies in the ready development of measurement techniques based on Endsley's (1995a) theory to be used in studies of system design or operator training.

Whereas Endsley's (1995a) model clearly delineates SA as a product, and its associated measurement techniques (e.g. SAGAT) consequently rely on retrospective memory-probes to assess the state of this knowledge, Smith and Hancock (1995) 'expressly deny that SA is merely a snapshot of the agent's current mental model' (p. 142). In their model, SA is neither product nor process but 'an adaptive capacity to guide behaviour in response to dynamic situations' (p. 142). Smith and Hancock (1995) illustrate this with the task of conflict detection in Air Traffic Control (ATC). Collision avoidance is a (obvious) key goal mandated by the regulatory authority. Horizontal separation and relative velocity are the two parameters that must be known in order to define the risk of collision. Combinations of values on these parameters define a 'risk space' that provides the necessary information for action. Smith and Hancock (1995, p. 144) propose that 'The risk space stands as an operational definition of SA' in this environment. Therefore, SA is not a mental model but an ecological specification of environmental constraints on goal attainment.

Cognitive Engineering

The work of Danish electrical engineer Jens Rasmussen has been highly influential in the newly developed field of cognitive engineering (Hollnagel & Woods, 2005; Woods & Roth, 1988). Characterized by concerns with the cognitive demands of work in complex sociotechnical systems, cognitive engineering tends to favour naturalistic studies of experts performing real-world tasks over laboratory studies of constrained or artificial simplifications of real-world problems. Laboratory studies of diagnostic reasoning have tended to examine cue utilization in the judgement of simplified (often written) descriptions of complex tasks such as financial or medical judgement (Brehmer, 1981).

Rasmussen (1993) argued that diagnostic reasoning in complex real-world domains was really a part of decision making and action and should be seen in context. In contrast to laboratory studies where participants are often provided with succinct input (e.g. values on a number of attributes) from which a judgement is to be reached, real workers are immersed in an environment from which they must extract relevant information and undertake appropriate action. Diagnosis and decision making are steps along a route, linking the environment to action rather than discrete activities.

Rasmussen (1986, 1993) described a 'decision ladder' (see Figure 2.2) which is proposed as a 'normative sequence' representing the relationships between states of knowledge. The performance of operators in real diagnostic tasks can, however, be represented on the decision ladder. For example, the recognition of a simple cue or pattern of cues may directly activate the execution of a standard response. This pathway directly links the lower levels (activation and execution) of the ladder: 'Experts in action ... have a repertoire of heuristic short-cuts by-passing the higher levels of the ladder' (1993, p. 982). The normal operation of a power plant, process control, ship or aircraft involves almost continuous low-level monitoring and activation to keep the system within desired parameters. This is what Rasmussen (1983) describes as skill-based control and is characteristic of most experienced operator actions when controlling a system in a steady state.

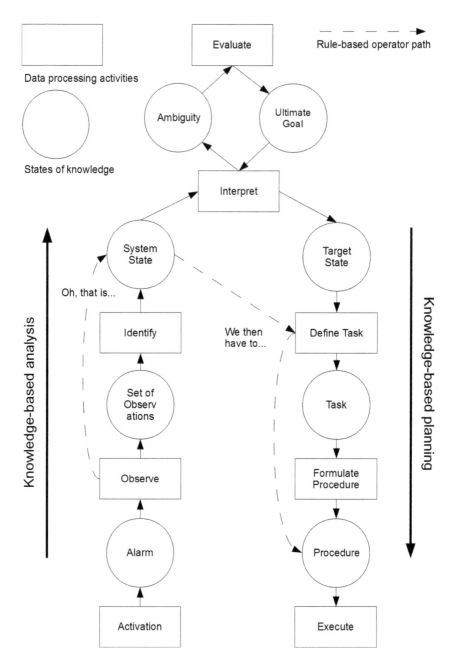

Figure 2.2   Rasmussen's decision ladder (adapted from Rasmussen, 1986)

If a signal or alarm suggests a non-normal state or condition, then one possible operator path is illustrated in Figure 2.2. Here, observation leads to an identification of a system state that requires a corresponding action. The operator's level of control has lifted a notch from 'skill-based' to 'rule-based'. The appropriate rules are instilled during training and the keys to successful action are the correct identification of the system state and the retrieval of the corresponding rule. The rejected takeoff scenario facing a pilot described at the beginning of the chapter provides a good example of rule-based control – there are several alternative system states depending on runway length and aircraft speed. Each has an appropriate response.

If the system state cannot be identified readily from the pattern of available cues, then operators may have to resort to functional reasoning based on their understanding of system dynamics. This leads to the higher levels of the ladder and the determination of appropriate goal-states and the strategies needed to bring them about.

Rasmussen's explanations nicely complement the perspective on decision making developed by Gary Klein (1989, 2008). Based on work with experts in the field, Klein (1989) noted that expertise in action was largely based on recognition-driven retrieval of previously acquired action patterns, rather than any kind of laborious problem-solving *in situ*. Experienced problem solvers develop a store of patterns – schematic structures incorporating relevant cues, expectancies and typical reactions. Therefore, the art of decision making, according to Klein (1989), lies in the ability to categorise or classify a situation. Once this is complete, the most promising response is automatically retrieved. Klein proposes further steps involving running a form of mental simulation to check that the typical response would actually be appropriate in the current circumstances. Klein (2008) has characterized this Recognition-Primed Decision (RPD) model as a blend of intuition (pattern recognition) and analysis (mental simulation).

Conclusion

Situational Awareness and diagnosis can both be seen as a process (e.g. 'we need to do some tests') undertaken to provide an answer to the question 'what's going on here?'. They can also both be seen as the product or outcome of that process (e.g. 'you have disease X'). Both terms are widely used in human factors research with 'situation' (or more correctly 'situational' awareness) predominating. Diagnosis and related terms such as 'diagnostic reasoning' are predominant in medicine and related fields. Situational Awareness has found wide acceptance in human factors engineering where technological advances have transformed the role of the human operator from 'hands-on' operator to more of a systems monitor. Maintaining an accurate state of awareness of the state of the system is naturally an important goal for the human operator. In contrast, despite huge technological advances in

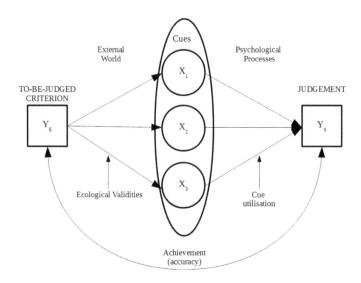

Figure 2.3   Lens model of judgement (adapted from Brunswik, 1956)

medicine, the human operator still retains very much a 'hands-on' role as assessor, interpreter and active planner of patient care. The process of diagnosis reflects a more active process of inquiry or truth-seeking in response to a problem (reactive). The process of Situational Awareness, on the other hand, has connotations of a steady state of maintaining a picture of an ongoing state of affairs (proactive).

Most models of diagnosis, as well as the more ecological models of SA, lay heavy emphasis on an analysis of cues in the environment. Making appropriate use of valid cues provides the key to effective performance in many dynamic environments. In this context, the validity of a cue (sometimes known as its ecological validity) refers to the correlation between a cue and a criterion (Plessner, Schweizer, Brand & O'Hare, 2009). These terms were first formulated into a coherent framework for understanding judgement under uncertainty by Egon Brunswik (Brunswik, Hammond & Stewart, 2000). Brunswik's 'lens model' (see Figure 2.3) of judgement, thus, underlies most modern work on the use of cues in diagnostic judgement.

The right-hand side of the model represents psychological studies of judgement in which attempts are made (usually by linear regression models) to capture the structure of human judgements. The left-hand side of the model represents the equally important but less studied question of the ecological validity of the available cues. For example, studies of medical diagnosis typically measure the weights that physicians attach to the various cues normally associated with a particular condition, although the ecological validities of these cues are rarely examined explicitly (Wigton, 1996).

Studies of diagnostic reasoning and performance in complex dynamic environments often use Subject Matter Experts (SMEs) to provide insight into the potentially valid cues in a domain. This method does not guarantee ecological validity as it is possible for experts to agree on the importance of a cue even though it may lack actual ecological validity, or to omit from consideration valid cues of which they are unaware. Given enough experts with sufficient experience in a domain then the likelihood of either of these problems can be considered minimal.

There is no doubt that effective performance in real-world domains begins with the detection of relevant cues and an understanding of their significance. These constitute the first steps on Rasmussen's (1983) decision ladder and the first two levels of Endsley's (1995a) description of Situational Awareness. Effective performance is predicated on addressing the actual problem. Studies of human error in aircraft accidents (e.g. O'Hare, Wiggins, Batt & Morrison, 1994) have shown that errors at the early stages (e.g. detection and diagnosis) are associated with more severe outcomes than errors at later stages (e.g. incorrectly performing a procedure or action).

Given this, it makes sense to target cue sensitivity and diagnostic understanding in training, especially for operators in time-critical or safety-critical domains. Wiggins and O'Hare (1995) found significant differences in search patterns and cue utilization between experienced and inexperienced pilots. Wiggins and O'Hare (2003a) developed a cue-based training programme for general aviation (GA) pilots. The cues were derived from cognitive interviews with a number of expert pilots. The software training tool provided participants with practice at recognising and responding to these cues as well as practice at incorporating these cues into decisions about continuing a flight under Visual Flight Rules (VFR). Evaluation of the training showed positive effects on both self-report and performance-based measures.

A similar logic has been applied to the training of performance in sports officials such as football referees (Plessner et al., 2009). Drawing explicitly on the logic of Brunswik's lens model, the authors set up a multiple-cue training environment involving numerous video clips of player contact situations in football matches. The fundamental task of the match referee is to correctly categorise each contact as a foul or as a non-foul. Criterion judgements were provided by a group of elite referees from the national football association. Immediate feedback was provided to participants after each judgement. Preliminary evaluations indicate that judgements can be improved by experience in an environment that provides repeated opportunities to learn the correct cue-criterion relationships through immediate feedback.

The ability to make use of appropriate cues has recently been developed into an approach to measure and evaluate expert performance (Wiggins, Harris, Loveday, & O'Hare, 2010). This set of cue-based tasks has successfully differentiated competent and expert diagnostic performance in a range of domains, including power control and paediatrics (Loveday, Wiggins, Searle, Festa, & Schell, 2013; Loveday, Wiggins, Harris, Smith, & O'Hare, 2013).

Neither diagnosis nor Situational Awareness can be considered explanatory or causal concepts (Flach, 1995). Both are purely descriptive. As noted earlier, there is a strong temptation to misuse the concept of Situational Awareness as a causal agent leading to unfortunate circularity of reasoning as pointed out by Flach (1995). Both diagnosis and more ecological models of Situational Awareness (e.g. Smith & Hancock, 1995) emphasise the importance of cues in modulating performance in dynamic systems. A focus on understanding cue-based judgement has led to innovative advances in the training and assessment of expertise. Continuing efforts in these directions should continue to bear fruit.

Chapter 3

Communication and Diagnostic Cues

Lidija Krebs-Lazendic, Nan Xu Rattanasone, and Jaime Auton

Clear communication is critical in maintaining safety in high-risk, high-consequence industries, such as air traffic control, rail control, and high voltage electricity transmission. At its core, effective communication requires that the listener understands the intentions of the speaker. This task can be difficult in safety-critical environments, which are characterized by ill-structured problems, high workload, and time pressure (Orasanu & Connolly, 1993).

Even during the most mundane communicative exchanges, there is the potential that the intentions of the interlocutor (i.e., a communicative participant) will be misunderstood. However, the causes of these misunderstandings are relatively easy to identify and the communicational flow is usually restored. Both the speaker and listener in everyday face-to-face exchanges of information are able to constantly monitor each other's responses for evidence of mutual understanding (Clark & Brennan, 1991) and to 'repair' the communication if such evidence is not forthcoming.

The difference between exchanges that take place within these everyday contexts and those that transpire within high-risk technical environments is that, in technical environments, the constant monitoring of interlocutors' utterances for evidence of mutual understanding is not always possible. It is not surprising, then, that in these environments, communication breakdowns are frequent contributors to inefficiencies and even fatal errors (Carthey & Reason, 2001; de Leval, Carthey, Wright, Farewell, & Reason, 2000; Sträter, 2003).

These communication problems can be divided into two groups:

1. those that are not strictly linguistic (e.g., a failure to provide the necessary information) or are associated with various technical issues (for example, a poor microphone or telephone transmission, or frequency congestion in air traffic control and rail traffic control) which result in a degraded speech signal; and
2. those that are strictly linguistic.

Although serious, the errors in communication that are not strictly linguistic can be mitigated through improvements in training. For example, operators who are trained to understand speech in noise (Song, Skoe, Banai, & Kraus, 2012) or to communicate effectively in specific organizational environments (Wouda & van de Wiel, 2013) have been shown to detect errors earlier and avoid miscommunication.

Of greater concern (and more difficult to solve) are the problems that arise from the complex interplay of language structure, language processing, and the linguistic and cultural backgrounds of the interlocutors. In this chapter, we describe the effects of ineffective and inaccurate communication on diagnosis, and how this impacts performance in advanced technology environments such as aviation.

Communication Breakdowns in Aviation: Causes and Consequences

Communication in the aviation context often involves operators from cross-cultural and multi-lingual backgrounds (Sullivan & Girginer, 2002). A particular area of concern is the miscommunication that takes place between pilots and air traffic controllers. Such miscommunication has been identified as a primary factor in several major aviation accidents (Cushing, 1997; Tajima, 2004), airspace incidents, and runway incursions (Boschen & Jones, 2004), as well as pilots' operational errors (Prinzo, 1996). For example, on 25 January 1990, the flight crew of Avianca Flight 55 failed to clearly declare a fuel emergency, resulting in fuel exhaustion and a collision with terrain. Similarly, on 26 September 2006, the flight crew of Gol Transportes Aéreos Flight 1907 misunderstood clearance instructions from air traffic control, resulting in a mid-air collision (Boschen & Jones, 2004; Cushing, 1997; Gil-Alana, Barros, & de Araujo, 2012; Helmreich, 1994).

One of the worst accidents in aviation history was a result of relatively simple miscommunication. On 27 March 1977, two Boeing 747s collided on the runway of Los Rodeo Airport in Tenerife, the Canary Islands, killing 583 and injuring 61 people (Bertrams, 1979). This accident is commonly referred to as the 'Tenerife' tragedy. The critical cause of the accident was a miscommunication between the Dutch-speaking captain and the Spanish-speaking air traffic controller.

Prior to the accident at Tenerife, both aircraft (Dutch KLM Flight 4805 and American Pan Am Flight 1736) were parked on the taxiway. When the aircraft began to taxi onto the runway, visibility was low due to dense fog. As a consequence, the pilots could not see the other aircraft, nor could the controller in the tower identify the positions of the aircraft on the runway. After a period of communication with the air traffic controller, the KLM flight attempted to take off while the Pan Am flight was still on the runway. The resulting collision destroyed both aircraft. The subsequent accident investigation revealed that, due to a communication misunderstanding, the Dutch captain believed that he had been given clearance to take off when no clearance had, in fact, been issued.

The Tenerife accident is a potent example of how a single word or phrase can be replete with ambiguity, even within a relatively specific communicational context. In this case, the word 'takeoff' was used incorrectly by the Dutch pilot and then misinterpreted by the air traffic controller. As the aircraft began to progress down the runway, the Dutch pilot told the air traffic controller: 'We are now at takeoff' instead of the unambiguous 'We are now taking off'. The air traffic controller interpreted the statement to mean that the airplane was in a stationary position at the takeoff point, waiting for clearance to take off.

The misuse and misunderstanding of the word 'takeoff' in this communicative exchange is an example of the pilot's inadvertent use of a Dutch grammatical construction with English words. In Dutch, the meaning of the English form verb + -ing (such as take + -ing in taking) is expressed as a + noun (such as at + takeoff). The Spanish air traffic controller, fluent in English but unfamiliar with Dutch, interpreted the structure at + takeoff in English as an indication of a takeoff point and failed to inform the Dutch pilot that the Pan American Boeing was on the runway.

The accident had a profound impact on the field of aviation communication, leading to the introduction of a strict communication protocol during takeoff, landing and taxiing. The protocol includes the mandatory use of standardized phraseology by both pilots and controllers, with the exclusion of the word 'takeoff' from any context other than a clearance to take off. However, even after the introduction of a protocol-driven communication, communication errors continue to cause misunderstandings (Prinzo & Britton, 1993).

In the following sections we will show how the complex processes of everyday communication, effectively due to the joint effort of interlocutors to understand each other, become fragile, ineffective, and error-prone in artificial/ non-typical communication contexts such as aviation. We will then explain how the likelihood of communication breakdown is greater in these contexts due to inherent constraints, including the lack of access to visual cues during telephone communication, which forces interlocutors to rely on prosodic cues, such as intonation, rhythm, and stress, to understand an utterance. Finally, we will discuss the impact of accented speech and language proficiency for non-native speakers of English on communication and, ultimately, on diagnosis.

Communication, Misunderstanding, and Non-Understanding

To communicate successfully, interlocutors exchange utterances collaboratively with the aim of achieving mutual understanding or 'common ground' (Clark, 1996; Clark & Schaefer, 1987; Clark & Wilkes-Gibbs, 1986; Cohen & Levesque, 1990). In other words, interlocutors in communication have a joint commitment to understand each other. The mutual understanding is achieved when all of the communicative participants comprehend information contained within the utterances and make inferences about goals that they believe others to be pursuing (Carberry, 1990; Cohen, Morgan, & Pollack, 1990).

The communicative exchange between the speaker and the listener contains two stages:

1. the presentation of the content; and
2. the grounding of the content (i.e., mutual understanding of the content).

Grounding is accomplished when the listener provides evidence that the speakers' intentions have been understood, and the speaker accepts the evidence as valid. Mutual acceptance typically leads to the continuation of the information exchange. If listeners are not able to identify the meaning of the utterance, they will signal to the speaker to expand the expression by adding further explanations or to replace the utterance with a new expression. The listeners then assess the meaning of the new utterance and either accept it (if the meaning is clear), or reject it again (if the meaning is not clear).

This communicative exchange continues until the referring expression is sufficiently acceptable for both participants to achieve mutual understanding. This final, accepted expression becomes the interlocutors' common ground and it further guides communication (Hanna, Tanenhaus, & Trueswell, 2003). However, given that referential meanings (i.e., the literal, specific, intended meaning) of utterances vary in different contexts, effective communication requires that listeners understand both the context-specific referential meaning of the utterance and the linguistic meaning of the language structure used in the utterance. If a listener fails to accurately identify the intentions underlying a speaker's utterances, the result is either a misunderstanding or non-understanding (Verdonik, 2010; Weigand, 1999).

During a misunderstanding, listeners fail to identify the intended meaning of the utterance, while also believing that their interpretation of the utterance is complete and accurate, thereby accepting the utterance (Hirst, McRoy, Heeman, Edmonds, & Horton, 1994; Skantze, 2005). Therefore, during a misunderstanding, neither communication partner is aware, at least initially, that an error has occurred. It is possible that a misunderstanding remains unnoticed in a communicative exchange and the participants continue to talk falsely, believing that grounding has been achieved. Alternatively, the conversation might break down, leading the interlocutors to determine that a misunderstanding has occurred.

A non-understanding occurs when the listener fails to find any complete interpretation of the utterance or is unable to choose between two or more alternative meanings of an ambiguous utterance (Skantze, 2005). In a typical non-understanding, the listener can immediately recognize the failure in communication (Hirst et al., 1994) and signals to the speaker that the flow of information has been interrupted, thereby leading the speaker to begin the repair process. However, in many cases, this does not happen. While not widely discussed in the literature, another possibility is that the listener conceals their non-understanding to save face. As a consequence, the speaker cannot identify that a problem has occurred.

In typical casual and low-risk communication contexts, this kind of elaborate reference negotiation (which normally includes multiple aspects and multiple stages of negotiation and repair of the problematic utterance) is very common (Bazzanella & Damiano, 1999). However, due to the severe time constraints experienced in many high-risk environments, interlocutors are often prevented from engaging in all collaboration stages in order to achieve mutual understanding. The absence of collaboration during a communication process has the potential to cause discrepancy between the speaker's intended meaning and the listener's interpretation of that meaning (Grice, 1989). This leads to errors in communication, which are often related to the unique nature of the communication environment (Kim, Park, Han, & Kim, 2010) and are less likely to be detected and repaired.

Communication errors are frequent, even when the interlocutors are speaking face-to-face, share a common language and cultural background, and have a common knowledge of the subject of the communication. However, communication breakdowns are more likely to occur when the interlocutors do not have access to visual cues, and/or are communicating in their non-native languages (i.e., second language or L2) or non-native dialect (Drury & Ma, 2002), as is often the case in complex technical environments.

Communications Protocols: Accident Prevention

To reduce the risk of errors in communication, a number of industries have introduced strict protocol-driven preventative strategies (Wadhera, Parker, Burkhart, Greason, Neal, Levenick, Wiehmann, & Sundt, 2010). For example, to reduce communication breakdowns in military and commercial aviation, a 'sterile cockpit' environment has been introduced during periods of high risk and high mental workload. These periods involve activities such as taxiing, taking off, and landing, as it is during these periods that pilots need to attend to communication with the controllers. During these phases of the flight, aircraft are in a critical transition stage, as they are either reducing altitude or speed to prepare for landing, or increasing altitude and speed to prepare for departure from the airport. These phases of flight are also the periods during which pilots experience the greatest workload, with landing the period during which there is the greatest risk of error and/or communication (Goode, 2003). Coincidentally, it is during takeoff and landing where voice transmission between pilots and air traffic controllers is both most frequent and most critical.

During takeoff and landing, all nonessential activities and conversations are prohibited. Pilots and air traffic controllers are expected to follow the standardized communication readback/hearback protocol using standardized phraseology (Australian Transport Safety Bureau, 2007). The purpose of standardization is to avoid semantic ambiguity and to expedite the communication process. The readback/hearback protocol is a radio or telephone procedure where the receiver of an instruction is required to repeat the instruction back to the sender (Streitenberger, Breen-Reid, & Harris, 2006). The protocol consists of four stages:

1. the sender delivers a verbal instruction;
2. the receiver actively listens to the instruction;
3. the receiver repeats the instructions to the sender (i.e., the readback stage); and

4. the sender actively listens for an accurate readback from the receiver (i.e., the hearback stage).

The purpose of this protocol is for the sender to confirm that the receiver has accurately heard the instruction (Schneider, Healy, & Barshi, 2004; Streitenberger et al., 2006).

The strict 'sterile cockpit' procedure, combined with the readback/hearback protocol, has been associated with a reduction in the frequency of error (Wadhera et al., 2010). This is due to the adoption of a specific conventional form of communication within a restricted range of dialogue options (McInnes & Attwater, 2004). This type of dialogue, known as the 'data transfer dialogue', is characterized by the precisely defined and distinct roles of *speaker* (i.e., 'data giver') and *listener* (i.e., 'data receiver'). The constrained nature of the data transfer dialogue, and the strictly defined roles of participants in the communication exchange, facilitates rapid and more effective mutual understanding, and reduces the likelihood of misunderstandings or non-understandings.

Despite a well-established system of conventions to facilitate mutual understanding, eliminating communication errors in aviation and other high-risk environments remains a challenge (Morrow, Rodvold, & Lee, 1994). The restrictive nature of protocol-driven communication does not allow the sender to assess whether the receiver has understood the meaning of an accurately delivered readback, since the ability to correctly repeat a string of words does not necessarily reflect an understanding of those words (Jones, 2003).

One reason for non-understandings in these contexts is the absence of visual cues (such as facial expressions and body gestures). Such cues typically contribute to the meaning of utterances, but they are not available to the sender in the strictly oral exchange of spoken information that occurs in readback/hearback communication. Hence, the sender is required to use prosodic cues encoded in the voice of the receiver (such as speech rate, stress, intonation, and pauses), combined with the linguistic elements of speech utterance (the syllables, words, and phrases), to detect a listener's non-understanding (Nygaard, Herold, & Namy, 2009; Nygaard & Queen, 2008).

Hearing but Not Seeing Speech: Prosodic Cues as Indicators of Non-Understanding

During face-to-face communication, visual cues, including eye gazes and various facial expressions, are crucial elements of the communication exchange. They are used to signal understanding or non-understanding (Drummond & Hopper, 1991), contribute to the meaning of spoken utterances (Krahmer & Swerts, 2005), and indicate the emotional states of interlocutors (Smith & Clark, 1993). Visual cues also help interlocutors structure their interaction by facilitating turn-taking and the rate of speech, thereby providing support for the information that the speaker wants to convey, and helping interlocutors structure the interaction accordingly (Clark & Krych, 2004). Visual information can also help speech comprehension in adverse acoustic conditions (Jordan & Sergeant, 2000).

During face-to-face communication, interlocutors achieve mutual understanding by combining the semantic cues of an utterance (i.e., identifying *what* has been said), and prosodic cues of an utterance (i.e., assessing *how* it has been said) (Ladd, 1996), into a single meaningful unit. This process of combining semantic and prosodic cues to identify the meaning of utterances is guided by visual information in the form of facial expressions (Krahmer & Swerts, 2005). For example, statements and questions can be distinguished via both intonation (e.g., falling voice at the end of a statement versus rising voice at the end of a question) and gestural response (e.g., flat eyebrows associated with statements versus rising eyebrows associated with questions) (Srinivasan & Massaro, 2003). Without changes in voice intonation and facial expression, spoken utterances such as '*Nine thirty one is ready*' (statement) and '*Nine thirty one is ready?*' (question) have identical meanings.

In the unique readback/hearback situation in which the receiver of an instruction is required to readback the instruction to the sender, the sender's utterance is simply restated. This does not provide any evidence of the receiver's understanding of that utterance. From a possible pool of cues, such as semantic, visual, and prosodic, the sender in an aviation readback/hearback context can only effectively use prosodic cues as evidence of the receiver's understanding or non-understanding during communication. Non-understanding arises when there is a perceived uncertainty about the meaning of the utterance, since uncertain information is considered incomplete, ambiguous, erroneous, and/or imprecise (Woods, 1988). Therefore, the sender's failure to detect uncertainty in the receiver's readback can potentially prevent mutual understanding and can cause breakdowns in communication.

Uncertainty is usually indicated through the use of linguistic hedges, such as '*I guess*' and '*I think*', a variety of fillers (e.g., '*hm*', '*oh*', '*um*'), as well as prosodic cues (i.e., long delays and rising, question-like intonation) (Smith & Clark, 1993). Listeners are sensitive to such prosodic cues and use them to make adequate estimates of the certainty or uncertainty of the speaker (Brennan & Williams, 1995). However, during face-to-face communication, speakers also signal their uncertainty visually, with appropriate facial gestures (such as rising eyebrows). Combined audio-visual signaling leads to listeners' more accurate perception of the speaker's uncertainty (Krahmer & Swerts, 2005; Swerts & Krahmer, 2005).

In the context of the readback/hearback protocol, the receiver of an instruction may signal understanding of the sender's utterance by repeating its semantic content correctly, but may indicate non-understanding using a rising, question-like intonation and filled inter-turn delays (Auton, Wiggins, Searle, Loveday, & Xu Rattanasone, 2013), combined with a rising of the eyebrows. Since the visual information (i.e., rising eyebrows) is unavailable to the sender, the sender is required to identify a non-understanding by relying solely on the prosodic cues of the utterance. Therefore, the failure to detect the inter-turn delays and question-like intonation, as an indication of non-understanding, may falsely lead the sender to interpret a semantically accurate readback as confirmation of the receivers' understanding of the instructions when, in fact, the accurate readback conceals non-understanding.

'World Englishes' in the Flying Tower of Babel: Linguistic Diversity and Communication Errors

Although the absence of visual cues in the readback/hearback protocol places particular weight on prosodic information during communication between pilots and air traffic controllers during the readback phase, misunderstandings can arise from linguistic structures at all levels of language (Bazzanella & Damiano, 1999; Gibson, Megaw, Young, & Lowe, 2006). These levels are:

1. phonology – language sound patterns;
2. syntax – language word patterns and sentence structure;
3. semantics – language meaning patterns; and
4. pragmatics – language in context (i.e., situational influence on meaning).

A serious contributor to air traffic communication breakdowns is the fact that many pilots and air traffic controllers speak English as their second (L2) or non-native language. This means that aviation communication is conducted in a linguistically diverse environment in which native English speakers are required to engage with different pronunciations of English. At the same time, L2-English speakers are expected to perform a number of cognitively demanding tasks in a high-risk environment, while also communicating effectively in their non-native language. Even with proficiency in English, unusual and/or high workload conditions may reduce non-native speakers' capacity to communicate and comprehend English utterances (Tajima, 2004). Misunderstandings in pilot–air traffic control communications are particularly likely when both the pilot and the air traffic controller are non-native speakers of English (Tajima, 2004).

It is typical for pilots and air traffic controllers to acquire English during their initial training. Spoken L2 fluency is very difficult to acquire when learning in adulthood. This is evident in all domains of language acquisition, but is especially pervasive in the pronunciation of L2 speech segments (i.e., vowels and consonants; Piske, Flege, MacKay, & Meador, 2002; Flege, Munro, & MacKay, 1995). Challenges with English pronunciation, perceived as accented speech by native speakers, are the result of previous linguistic experience with the first (L1) language (Flege, 1995). Late learners show stronger L1 effects than learners who learnt their L2 in early childhood. Because each L1 influences subsequent L2 learning differently, adult speakers not only acquire inaccurate, accented pronunciations of their L2, but these accented pronunciations are L1-specific (for example, L1 English-speakers can distinguish between 'Japanese'-accented English, 'Spanish'-accented English, 'Arabic'-accented English, etc.).

Unfamiliar accents are particularly difficult to understand for native speakers (Trude, Tremblay, & Brown-Schmidt, 2013) and may significantly impact their comprehension and responses in communication (Fegyveresi, 1997). On the other hand, for non-native speakers, the speech rate of native speakers is often perceived as too fast (Henley & Daly, 2004; Itokawa, 2000). Although the majority of L2 pilots and air traffic controllers are proficient in English, this may not be sufficient when coping with emergencies and having to communicate at a rapid speed with their attention divided between various tasks in order to make a time-constrained decision.

Despite the introduction of English as the international language of aviation, with mandatory testing of English for all non-native speakers of English, comprehension difficulties have continued to be cited as the primary cause of operational airspace incidents (Cushing, 1997). Since English training is usually conducted in the Standard English of the country of instruction, the majority of English-L2 pilots and controllers are not exposed to the idiosyncratic characteristics of different English dialects and different foreign-accented English pronunciations during their training. However, most pilots and controllers will encounter non-familiar accented English in the linguistically and culturally diverse aviation communication context during the course of their career. This lack of experience with various non-standard English pronunciations may affect their understanding of critical information, thereby increasing the risk of miscommunication.

Air traffic controllers also frequently communicate with pilots who speak accented English and must rapidly adapt to perceiving and understanding foreign accented utterances. This means that both pilots and controllers are expected to be able to interpret utterances from different speakers, with large variabilities in speech rate, prosody, and accents. Similarly, they must also tune their perception to ensure that they understand the meaning of a specific utterance, while ignoring all other uninformative and potentially confusing variations contained in the production of that utterance.

Pilots learning Standard English acquire prosodic cues (such as intonation, rhythm, and pause) typical for conversational English. However, the use of prosodic cues significantly differs between standard conversational English and aviation communication. In aviation English, the speech rate is faster, while intonation and pauses, used in everyday communication, are often missing (Prinzo, 2008). In addition, for the speakers of languages that place more importance on visual cues than does English (e.g., Catalan and Dutch; Sendra, Kaland, Swerts, & Prieto, 2013), the absence of visual cues may result in greater difficulty in communication than that experienced by native English speakers. While the strict use of jargon and aviation phraseology in specific contexts reduces the likelihood of misunderstandings, the rapid speed at which controllers deliver instructions may still be problematic for non-native speakers, who tend to be more reliant on the pronunciation than on the semantic context when comprehending speech in L2 (International Civil Aviation Organization, 2009). Finally, training in the use of the specific jargon and phraseology may reduce English-L2 speakers' ability to effectively use a larger vocabulary in English. This is especially important in the case of unusual situations and emergencies, when non-standard English idioms are likely to slip into the communication and make the situation much more difficult for English-L2 pilots and controllers.

Conclusion

While protocol-driven communication is a remarkable step towards reducing communication errors in cognitively-demanding environments (Wadhera et al., 2010), it is certainly not sufficient to eliminate potentially fatal communication errors. However, the adherence to communication protocols and standardized phraseology in many high-risk environments can benefit from adaptation to non-familiar dialects, as the constancy of relevant context and rigid syntactic structures may facilitate the understanding of non-familiar pronunciations (Yoon & Brown-Schmidt, 2013). In addition, exposure to dialectal and accented speech as part of initial training may also improve operational outcomes. Research in psycholinguistics has shown that despite initial challenges with impaired perception of unfamiliar accented speech, performance improves with repeated exposure and, in some cases, matches the performance of native speakers (Bradlow & Bent, 2008; Clarke & Garrett, 2004; Sidaras, Alexander, & Nygaard, 2009). This occurs despite training within limited contexts and over short exposures (Witteman, Weber, & McQueen, 2013). This adaptation is equally possible for both different regional dialects from the listener's native language (Trude & Brown-Schmidt, 2012) and foreign-accented speech (Trude et al., 2013). Therefore, training L2 English-speaking pilots to perceive different types of accented English may reduce the likelihood of later misunderstandings in pilot–air traffic control communication, regardless of their native language(s).

Training pilots and air traffic controllers to encode prosodic cues from vocal expressions to identify non-understanding remains particularly challenging, since prosodic cues are often ambiguous without accompanying visual cues. However, it is likely that the strict readback/hearback phraseology can be beneficial for prosodic cue learning, in the same way that it is useful for training different accents. Since the phraseology in the protocol remains constant when speakers are in different emotional states, air traffic controllers can be trained to attend to changes in the prosodic cues associated with specific emotional states when assessing the level of understanding of an utterance. Cue-based training has been shown to improve the capacity of pilots to recognize hazardous weather conditions (Wiggins & O'Hare, 2003a), improve the ability of miners to avoid hazardous rock falls (Blignaut, 1979), and to decrease the response times of sports people (Abernethy, Wood, & Parks, 1999; O'Hare, Wiggins, Williams, & Wong, 1998; Smeeton, Williams, Hodges, & Ward, 2005). As a consequence, implementing training in the detection of prosodic cues, such as rising voice at the end of utterances and filled inter-turn delay, may increase the recognition of non-understanding of instructions, thereby improving communication-related diagnosis.

Chapter 4

Vigilance, Diagnosis, and its Impact on Operator Performance

William S. Helton

Diagnostic expertise is evolutionarily adaptive. While the focus of this book is on the role of diagnostic expertise in occupational settings amongst people, the ability to acquire expertise is widespread amongst animals (Helton, 2008). The present chapter will place diagnostic expertise in wider ecological and physiological context. In particular, in this chapter, the findings from the scientific literature will be discussed that are relevant to placing diagnostic expertise in evolutionary context. The neurophysiological limitations which sub-serve diagnostic capabilities will also be considered. The discussion will then focus on a key aspect of many ecologically realistic tasks that place a specific resource demand on operators: temporally extended searches. Finally, the role of sustained attention will be examined, particularly in understanding its limitations, as this provides a basis to improve operational performance.

Evolutionary Context

In the savannah of Africa, a cheetah eyes a potential meal. The gazelle springs through the tall grass, flexing and un-flexing its powerful leg muscles. The cheetah springs into action, becoming a blur of yellow. Quickly the cheetah stops. The cheetah's accurate diagnosis of the gazelle's ability to spring athletically through the grass meant that pursuit was folly. Caro's (1994) ethological research on cheetahs is one of the most extensive investigations of skill development in a non-human predator and provides evidence of the role of diagnostic skill in hunting performance. One of the primary differences that Caro has found between novice adolescent cheetahs and seasoned adult cheetahs is the average distance needed to abandon the pursuit of prey. Inexperienced cheetahs, on average, abandon a chase after 18 metres, whereas seasoned cheetahs abandon chase after only 2 metres. Experienced cheetahs appear to have more skill or a greater ability to detect the cues necessary for determining when a chase is futile. The seasoned cheetahs appear to have developed acute abilities to 'read' prey, and they can directly perceive when a chase will be a waste of their precious energy.

Interestingly, even predators with relatively limited nervous systems appear to be advantaged by the ability to decipher the cues of their prey. The relatively small jumping spider (Portia fimbriata) is a visual hunter of dangerous prey – other spiders. Indeed, Portia hunts spiders who are themselves hunters: Portia's prey is typically web-building spiders. Web-building spiders construct massive snares to trap their own prey. To hunt these web-building spiders, Portia has to risk becoming prey by attacking the spider on its 'own turf' – a relatively massive trap which serves as an additional perceptual sense for the web constructor (extending tactile perception via vibrations on the web). The most effective time for Portia to act is when the web-building spider is busy wrapping up its own prey. The vibrations generated while the web-building spider is wrapping its own prey serve, as Jackson, Pollard, and Cerveira (2002) express it, as a 'cognitive smokescreen'. The wrapping vibrations mask the vibrations of Portia moving on the web. Portia can kill a deadly predator in its own lethal trap because Portia can perceive the cues that indicate when the other spider is sufficiently distracted to be vulnerable to attack. Appropriate cue perception and diagnosis is the difference between life and death.

Animals are barraged with extensive, indeed vast, amounts of sensory information during their interactions with their inhabited environment. Most animals, especially complex vertebrates like people, have a staggering array of sensory receptors. Therefore, almost all animals are operating under conditions of sensory overload where the amount of information needing to be processed is daunting. For this sensory information to be useful in guiding the animal's actions, the information has to be processed, integrated, and filtered. This information processing is subject to resource limitations as the animal's biological computations require both energy input and the use of a finite amount of information-processing structures. The brain of, for example, a person is only 2 per cent of the person's body weight, but uses 20 per cent of the person's metabolic resources. All animals, including people, have to work under severe economic constraints. Processing all of the available sensory information all of the time would quickly overburden the central nervous system and require energy inputs that no living animal could actual meet. Brains, simply, are not cheap organs. Therefore, animal brains have found computational shortcuts and means to process information frugally. Psychologists and cognitive scientists who study this frugal processing simply label it as 'attention' and have divided it into four functions or facets: selective attention, focused attention, divided attention, and sustained attention (Davies, Mathews, Stammers, & Westerman, 2013).

Animals need a means of selecting which sensory channel or stream of sensory information will be given processing priority. This process is labelled selective attention (Davies et al., 2013). Once the channel or stream of information has been selected for further processing, the animal needs a mechanism to limit the processing of the non-selected or irrelevant channels or streams. This process is referred to as focused attention (Davies et al., 2013). Sometimes, the animal is forced to process two or more streams simultaneously. For example, a predatory spider, while focusing on a tasty prey, has to also monitor other sensory streams for the cues of other predators. This is labelled divided attention. Lastly, there are times when the animal may have to extend the temporal window within which it focuses a search for a particular cue or cues. The animal has to force itself to continue searching for the cues, even when there is little external support for the search (e.g. the targets occur very infrequently). This is termed sustained attention. The key is that these four types of attention are functional descriptions, and they are not necessarily conceptualized as distinct systems or processes built into the architecture of the animal's central nervous system.

Information processing limitations have, themselves, been a major constraint and a driver of animal evolution. For example, Dukas and Kamil (2000) found that the blue jay's search behaviour is impacted by processing limitations. The authors trained blue jays to perform a search for digital prey (virtual moths) presented on touch screen displays. When the prey displayed in the centre of the screen were highly cryptic (perceptually difficult to detect), the blue jays were less able to detect prey appearing in the peripheral parts of the screen. When the prey displayed in the centre of the screen were less cryptic, detection performance for peripheral targets improved. In a follow-up study, Dukas and Kamil (2001) trained blue jays to search for two distinct types of prey. When the blue jays divided their attention between the two prey types, they had impaired detection performance compared to the search for one particular type of prey. Divided attention conditions appeared to decrease the blue jays' foraging ability. In further studies, Bond and Kamil (2002) found that the blue jays' attention limitations could actually drive the evolution of their virtual digital moth prey. The digital moths were forced to become more cryptic and phenotypically diverse in response to the blue jays' search behaviour. As the moths stretched the blue jays' information processing resources, they ensured their own survival. However, limitations to attention not only drive the evolution of prey, it also drives the evolution of predators who may become increasingly specialist foragers to cope with the processing demands that are induced by increasingly phenotypically diverse and cryptic prey. The predators may encode or quickly acquire specific prey-search templates. In behavioural ecology, this is known as a search image. Animals' search images in the terminology of this book would be a set of associated cues for a particular type of prey. When the prey are cryptic and the cues are initially difficult to perceive, the animal cannot monitor for multiple prey types, as the range of cues for which the animal has to divide its attention becomes quickly overwhelming. The animal has no choice but to limit its own search behaviour.

In humans, the limitations of biological information processing is well understood. Everyone has had experience of not being able to concentrate fully on more than one task simultaneously. For example, when experienced drivers are driving, they can drive while simultaneously carrying on a conversation. However, should the driving conditions become challenging, such as on ice with reduced visibility, then even an experienced driver may be unable to carry on a conversation during the drive. 'Something' that the driver needs to perform the two tasks simultaneously is in limited supply and, as one of the tasks becomes highly demanding, the other suffers. Psychologists refer to this limited 'something' as the cognitive resource (Kahneman, 1973). When describing these resource limitations, psychologists often use metaphors to help other people understand the underlying process. The two most common metaphors used to explain cognitive resource limits are hydraulic or monetary metaphors.

In the hydraulic metaphor, the cognitive resource is considered a pool (or pools) of mental energy, analogous to the fuel used to power energy or power systems (Hirst & Kalmar, 1987). The metaphor should not be taken literally. No one believes there is an attention or cognitive 'goo' which is held in some central tank and passed to sub-engines of the brain to enable processing (although in modern brain imaging, blood is sometimes referred to as a resource). Conceptualizing a central pool or pools, however, allows the psychologist to envision how separate processes can demand a common source and how, when the resource pool is redirected to another process, this can negatively impact a competing process. This metaphor also makes it easy to visualize that the resources are finite and can be over-consumed, thereby leading to impaired performance (analogous to an automobile engine failing due to running out of fuel).

In the economic metaphor, resources are analogous to money and the allocation of resources is limited by the economics of supply and demand. The advantage of the economic metaphor (the brain's bank account) is that it is less conceptually limiting than the fuel analogy. Money is not gold nor any other physical entity. Money represents abstracted value. Similarly, cognitive resources may reflect anything of limited supply in the animal's central nervous system, including glucose-oxygen, neurotransmitters, neurons, or actual neuronal groups. If a person was to make a choice of building a new addition to a house or upgrading a kitchen, money is a relatively rapid means to represent the total resource limitations of the entire building system (supply of wood, nails, workers, architect time, etc.). Money is a good metaphor, as it allows psychologists to talk about resource limits in an understandable format without being forced into specifying the exact physical source of the limits (what 'stuff' is actually limited). Neither the economic nor the hydraulic metaphor should be taken literally. They are merely useful conceptualizations for helping people grasp resource limits in biological computation.

Animals can reduce the processing demands of cue detection with experience. In this case, the resource analogy would be the development of a more efficient process. Using a hydraulic metaphor, the animal may shift to a more fuel efficient process or, using an economic metaphor, the animal may have developed, with practice, a more time-efficient or material-efficient means to do the same task. The ability of animals, including people, to learn which sensory cues are reliable indicators of the presence of hard-to-detect or cryptic targets is critical (Wiggins & O'Hare 2003b; Wilson, Helton, & Wiggins, 2013). Merely exposing the sensory systems of animals to stimuli repeatedly appears to improve the information processing of those stimuli (Mukai, Kandy, Kesavabhotla, & Ungerleider, 2011; Watanabe et al., 2002). Nevertheless, mere exposure to stimuli repeatedly does not itself, in all cases, guarantee improved detection performance. While some studies do indicate that perceptual learning can occur even to unattended stimuli (subliminal learning; Watanabe et al., 2002, 2001), performance in other studies shows greater improvement or an accelerated rate of learning when the observers exhibit greater cognitive effort directed explicitly towards the stimuli (Mukai et al., 2007; Mukai et al., 2011). Essentially, increased cognitive effort is indicative of a top-down active or attentive search for relevant perceptual cues. This suggests that the initial cue learning processes is resource demanding, but that these demands reduce, although they may not be eliminated entirely, with practice or increased experience. Therefore, experts or specialists should have less continuous cognitive resource expenditure than novices (see Small, Wiggins, & Loveday, 2014).

Increased cognitive effort is even detectable at a neurophysiological level. For example, Mukai et al. (2007) noted in a functional magnetic resonance imaging (fMRI) study, higher levels of activity in the dorsal fronto-parietal regions of the brain in participants who had accelerated rates of perceptual learning. The increase in frontal brain activity was interpreted as an indicator of more executive or endogenously controlled attentive focus on the stimuli. Similarly, in a later study, Ong, Russell, and Helton (2013) employed functional near-infrared spectroscopy (fNIRS) measures of frontal cortical activity to evaluate the role of neurophysiologically determined cognitive effort in perceptual rates of learning. In their experiment, they had participants attempt to detect the presence of buried bodies in snow over multiple inspection blocks. While the overall group trend was for improved perceptual performance, the rate or slope of perceptual learning tended to be steepest for those participants with higher initial frontal activation. The initial frontal activation declined, however, with increased practice and experience. This suggests that the demands of the task reduced with increasing detection skill.

Sustained Attention

While the cognitive resource demands reduce with increased practice, searching for cryptic perceptual cues is still demanding, especially if the search continues over long temporal windows or across large spatial extents. As mentioned in the previous section, there are times when an animal may have to extend the temporal window in which it focuses a search for a particular cue or pattern of cues. The challenge is that the animal must force itself to continue searching for the cues, even when there is little external support for the search (e.g. the targets occur very infrequently). Psychologists label this process 'sustained attention' or 'vigilance'. The primary source of information on sustained attention in animals comes from research on humans (Helton et al., 2005; Helton & Warm, 2008), with a growing body of research on rats (Bushnell, Benignus, & Case, 2003) and some relatively new research on dolpins (Ridgway, Carder, Finneran, Keogh, Kamolnick, Todd, & Goldblatt, 2006; Ridgway, Keogh, Carder, Finneran, Kamolnick, Todd, & Goldblatt, 2009). Similarities in sustained attention performance for humans and rats have been identified, whereas dolphins appear to be somewhat unique due to their specialized unihemispheric sleep (e.g. literally one brain hemisphere can rest while the other works, thus ensuring sustained operations).

The term vigilance was originally used by Head (1920, 1923) to describe a state of maximal physiological and psychological readiness to react. While there was an older literature examining performance changes over time (Bills, 1935), research on vigilance did not become systematic until Mackworth (1948) began work on the temporal pattern of performance amongst radar (radio detection and ranging) operators for the Royal Air Force during the Second World War. Radar was a new technological development during the Second World War and provided combatants with a significant advantage. For example, the British deployed radar in the search for surfaced Axis submarines. The hunt for Axis submarines was of critical importance for the island nation and the British were quick to adapt the new technology in the hunt. The British would deploy aerial radar units, but despite being highly motivated, the radar operators would begin to miss the blimps on their screens (which could be surfaced Axis submarines) within 30 minutes of the start of a watch. Being keen to figure out what was happening, the Royal Air Force commissioned Norman Mackworth to conduct research in an attempt to solve the problem.

To experimentally study the phenomenon, Mackworth developed the Clock Test. In the Clock Test, the participant was assigned to watch the movement of a black pointer (clock hand) on a blank-faced circle. The normal movement of the pointer was 7.5mm occurring once per second. Occasionally (less than 5 per cent of the time), the pointer would make a 15mm movement (a double jump). When these double movements occurred the participant was to let the experimenter know by pressing a key. Despite the double movements being clearly discernible from the single movements, participants' ability to detect the double movements quickly waned with time on watch. After 30 minutes of watch, the participants' detection rates dropped 10 per cent and continued to drop with continuous time on task. In addition, the response time for participants to press the key after the double movement also increased with time on watch. This decline in performance efficiency is known as the vigilance decrement and is characterized as an increase in response time to critical signals, an increase in missed signals, or both. The vigilance decrement has been observed repeatedly in both laboratory and operational settings, and with a variety of temporal search tasks (Helton et al., 2005).

Unfortunately, the vigilance decrement and the underlying reality of limits of sustained attention in people place severe constraints on the benefits gleaned from increased automation in the workplace. Automation is intended to relieve people from engaging in demanding work and to reduce task-induced stress (Parasuraman & Wickens, 2008). However, one of the consequences of imperfect automation is to require the human operator to become an automation monitor. This assignment places intense demands on sustained attention. While earlier researchers argued that vigilance and monitoring duties, while boring, were low in overall demands, this turns out to be wrong (Warm & Dember, 1998). Vigilance and monitoring assignments are often subjectively boring, but they are also highly demanding and stress-inducing. Vigilance increases both self-reports of workload and stress, and physiological indicators of stess, including elevated amounts of circulating catecholamines and corticosteroids (biochemical markers of stress; Parasuraman, 1984). Prolonged monitoring assignments can also cause stress-related changes in heart rate and galvanic skin response, and an increase in muscle tension and muscle tremor (and fidgeting) (Galinsky, Rosa, Warm, & Dember, 1993; Davies & Parasuraman, 1982). These physiological results are matched by self-report measures in which people who perform vigilance assignments report elevated levels of distress and fatigue (Galinsky et al., 1993; Helton et al., 1999, 2005, 2008). Vigilance assignments are not at all easy nor benign. In fact, they can be excruciating for many people.

By taking the perspective that cognitive resources are limited, insights are provided into the operation of vigilance. Vigilance assignments require extended temporal searches. The activity of searching for targets requires effort and, therefore, cognitive resources. Since the task requires the same search activity, and the search is extended temporally, there is no possibility to relieve the search system of continuous demands. In a vigilance task, the operator needs to make continuous 'signal plus noise', versus simply 'noise' discriminations. Depending on the nature of the task, there is no time in which the operator is able not to search. In many real vigilance tasks, target occurrence is completely unpredictable. The targets are often cryptic. The targets often do not 'pop-out' at the operator, but require the operator to filter out noise and clutter. Depending on the perceptual difficulty (the objective psychophysical difficulty) of the task, the operator must continuously perform a challenging perceptual discrimination. Therefore, as the search continues temporally, the demands on cognitive resources outstrip supply. The central nervous system presumably also regulates this activity and may take protective actions to maintain some level of performance. As a result, it may regulate expenditure.

The resource theory perspective is more helpful in understanding and predicting vigilance performance than alternative perspectives which have sometimes gained more traction in the general public and even otherwise uninformed professionals. For example, many theories of the vigilance decrement do not take a limited resource theory perspective, but instead, consider the decrement to be a result of boredom induced by objective task monotony. These alternative theories contrast with resource-based theories by suggesting that the vigilance decrement is due to task under-load. These boredom or under-load perspectives have immense appeal as they seem to reflect many people's personal experiences in vigilance tasks or vigilance task phenomenology. Vigilance tasks are often subjectively boring. These tasks can often be objectively monotonous with relatively little change in stimulation over time. This belief is exacerbated by the kinds of laboratory tasks that are used to dissect vigilance, which are often very simple, like Mackworth's original clock task. A natural inclination is to believe that task monotony drives subjective boredom, and that performance suffers due to the organism's free decision to reduce effort on the task. This view, while enticing, has limitations. First, it fails when making predictions about the relationship between objective task difficulty, objective monotony, and performance (Helton & Warm, 2008). Second, it never considers why task monotony is aversive, why organisms experience boredom, and what function it might serve. Finally, under-load theories fail to consider the research implications of comparative work on vigilance.

In relation to the first issue, research has demonstrated several findings regarding the relationship between objective task difficulty and vigilance performance. For example, there are several effects that have been consistently replicated in vigilance research:

1. Signal Salience Effect: When signals are made more difficult to detect, by lowering the signals' contrast with their background environments, performance tends to reduce and the decrement is more pronounced (Helton & Warm, 2008).
2. Event Rate Effect: The more rapidly the signals plus noise and noise stimuli are presented, the more rapidly performance reduces and the decrement is more pronounced (see Howe, Warm, & Dember, 1995).
3. Spatial Uncertainty Effect: The more uncertain is the spatial location of the signals, the greater the decrement in performance (Helton, Weil, Middlemiss, & Sawers, 2010).
4. Temporal Uncertainty Effect (sometimes called the Signal Regularity Effect): The greater the uncertainty as to when signals will occur, the greater the decrement in performance (Helton et al., 2005).

Employing a boredom hypothesis, one would probably predict exactly the opposite to these findings. If the task is more challenging and a signal is more difficult to discriminate (lowering signal salience or intensity), increasing the event rate or the spatial and temporal uncertainty should result in lower levels of boredom and, thus, performance in more challenging scenarios should improve. However, from a resource theory perspective, any increase in the demands of the task will impede performance so that a more challenging task will eventually result in lower performance. This is exactly what occurs when performance is tested in more challenging tasks.

Monotony-based theorists rarely appear to attempt to modify the objective monotony of the task and determine whether this aspect of the task is a critical driver of the vigilance decrement. This is true both in the intrinsic monotony of the task where the nature of the task itself changes, and in extrinsic monotony, where there are changes in non-task related features, like adding secondary tasks. In a recent study, Head and Helton (2012) examined this issue by having participants complete a vigilance task consisting of non-repeating forest or urban picture stimuli as target stimuli. Despite the use of non-repetitive, natural picture stimuli as targets and neutral stimuli, a statistically significant vigilance decrement occurred in the

20-minute period of watch-keeping. This was somewhat surprising as the accurate recognition of forest from urban scenes can be made relatively quickly (Greene & Oliva, 2009). Further, natural-forest scenes tend to elicit inherent fascination and increased visual scanning (Berto, Massaccesi, & Pasini, 2008). The finding that even when using non-repetitive forest stimuli as targets (and the pictures were not degraded), there is still a vigilance decrement is challenging for under-load theories. While under-load theorists might argue that the task is still subjectively monotonous – 'once you have seen one forest picture, you have seen them all' – this would make the theory viciously circular as there would be no way to *a priori* predict what drives subjective monotony.

Another possibility to test monotony-based explanations of the vigilance decrement is to add a secondary task to the main vigilance task. By having the watch-keeper engage in another task, the goal would be to increase the overall load and thereby forestall the monotony-induced vigilance decrement. While an initial study was interpreted as supporting this proposition (Ariga & Lleras, 2011), subsequent studies have clarified the impact secondary tasks have on vigilance (Helton & Russell, 2011, 2012, 2013; Ross, Russell, & Helton, 2014). If the secondary task enables a rest from the vigil, it can be recuperative, but if it entails a similar processing load (overlapping resources) or a concurrent load with the vigilance task (parallel operation), secondary tasks negatively impact vigilance performance. These findings are more consistent with a resource-based perspective and are challenging to integrate with an under-load or boredom-based explanation.

Even if monotony was the driving mechanism, the question remains as to why task monotony is aversive. If an analogy between mental and physical workload is plausible, a similar question might be asked. Why, when running on a treadmill, does the runner eventually tire of the task and stop? If boredom is a signal, it is unclear what it is signalling and what evolutionary function it serves. In the case of physical workload, the fatigue signal is a self-preservation signal that prevents self-harm or a catastrophic shutdown (Noakes et al., 2005). It is unclear why mental workload would operate so differently from physical workload. Contemporary science rejects Cartesian dualism. Mental workload is a result of physical (brain) workload. Even if boredom was to induce task performance reduction, it remains unclear why this would occur. The answer would likely be to prevent damage or catastrophic resource loss, which is more severe when the task is continued, demanding, and allows for no rest-recovery.

Finally, under-load theorists do not consider the implications of comparative or cross-species research on the vigilance decrement. While there has not been extensive research on vigilance in other species, the research thus far indicates that rats, dogs, and people all suffer the vigilance decrement, but curiously, dolphins do not, even when their vigilance requirements extend over several days (Ridgway et al., 2006, 2009). This pattern, where terrestrial mammals have a vigilance decrement, but some aquatic mammals do not, is very difficult to explain from an under-load perspective. From a resource theory perspective, the finding is not especially puzzling, since dolphins have evolved the ability for uni-hemispheric sleep. One side of the brain sleeps, while the other works. Otherwise they would drown. The 'upside' for dolphins is that this means that their two hemispheres are presumably more independent and, therefore, they have essentially two cognitive resource pools from which to draw. Dolphins can work one brain system, let the other rest, and then switch. Unfortunately, terrestrial mammals, like people, do not have the adaptation for uni-hemispheric sleep. A problem with under-load theories is that they are not only trapped in mentalist concepts (boredom, mindlessness, etc.), but they have a pre-Darwinian perspective. Under-load theorists never consider how the vigilance decrement operates in other animal systems and how this may inform or completely discredit their underlying theory. While under-load theorists could posit that the vigilance decrement in people is inherently unique and that findings from other animal systems have no implications for the theory regarding people, this would be difficult to accept in post-Darwinian science without justification (meaning the burden of proof of non-relatedness rests with those claiming zero continuity).

Conclusion

Sustained attention is a concern in prolonged tasks requiring perceptual based diagnoses, in which the organism makes continuous signal plus noise versus noise discriminations. The best scientific explanation of the declines in performance noted in temporally extended search tasks (the vigilance decrement) is one based on resource limitations. Alternative theories of the vigilance decrement, while popular and believable given phenomenological reports, reduce the process to mentalistic nonsense based on non-evolutionary, non-naturalistic accounts of human mental function. The central nervous system is a physical machine and like all physical machines is resource limited and subject to physical degradation. The cause of the vigilance decrement has something to do with resource allocation and expenditure, unless the vigilance decrement is the result of a non-physical (as in not real or mumbo jumbo) process. A resource-based perspective does not, however, rule out the role that resource allocation plays in task performance. People may, indeed, choose to redirect their limited resources to tasks other than the vigilance task. However, even in this case, the decrement is the result of limited (redirected) resources. Allocation is an issue where resource theorists may eventually subsume some aspects highlighted by under-load theorists, but this should not cloud the issue that, in operational settings with motivated workers, the primary driver of the vigilance decrement is resource depletion.

The operational advantage of a resource perspective is that it suggests three very clear paths to improve vigilance performance:

1. Reduce the objective difficulty of the task;
2. Provide rest breaks; and/or

3. Improve the ability of the operator to perform the task in a resource conservative fashion.

The first path simply requires the design of the task to make the targets highly salient, presented at a reasonable pace (too slow would, of course, reduce the speed at which the task can be completed), with the signals occurring in as predictable times and places as possible. This is probably exactly the opposite of what the phenomenological boredom perspective would advise, but it definitely appears to work and is corroborated by extensive laboratory research. The second path involves the provision of rest breaks. These have also been found to improve vigilance performance (Ross et al., 2014). Third, vigilance operators can be trained to detect signals. Perceptual performance does improve with practice, presumably because the operator is finding more efficient ways to detect the signals. Anything that reduces resource demands will improve vigilance performance.

This is fundamentally where diagnostic expertise meets the vigilance decrement and where a resource theory perspective is informative. An under-load or boredom model would probably suggest the opposite: that highly skilled operators would get more bored and, thus, make more errors on vigilance tasks. Of course, this does not match the empirical findings, in which skilled operators have a reduced, not exacerbated, vigilance decrement (Small et al., 2014).

Chapter 5
Designing for Diagnostic Cues

Thomas Loveday

When Klein (1989) first postulated that cue-utilization was a core component of expertise and diagnostic performance, his observations were predominantly drawn from interviews with fire commanders. At time of the interviews, one of the characteristics of this work-role was that decision-making was performed locally. This meant that many of the features available to the fire commanders occurred in real-time and were directly linked to important outcomes. However, increasingly we find that this type of diagnostic task is the exception rather than the rule.

As the technology underpinning automation and computerization has improved, an increasing proportion of the skilled work force is now tasked with performing complex diagnosis from centralized control-room settings. The motivation behind the shift to remote monitoring is often framed in terms of economic efficiencies, mostly because centralized control centres allow fewer operators to monitor a larger proportion of the system. However, practitioner preferences are also a contributing factor. After all, remote monitoring has made many jobs easier with new, centralized stations allowing multiple systems to be monitored simultaneously.

Although the transition from local to remote monitoring has increased the efficiency of many systems, it has also changed the nature of the features available to the operator during diagnostic tasks. Remote environments and interfaces do not allow direct observations of features and their related outcomes, from which practitioners have traditionally built a repertoire of naturalistic cue associations. Instead, in remote environments, practitioners must rely on artificially constructed representations to prompt their recognition that a critical event has occurred, and to communicate the cause and outcomes of that event.

As this chapter will illustrate, a key component of good system design is the integration of features that cue the recognition of a system's status accurately and efficiently. Further, this chapter will argue that system features can be designed to cater to the operator's capacity to use cues effectively. The chapter will begin with the discussion of several case studies in which operators were unable to recognise the status of the system due to poor feature design and implementation. These case studies will be considered from a recognition-primed decision-making perspective, with a particular focus on how the relevant design issues were resolved. The chapter will then transition to a more proscriptive approach, discussing how the recognition-primed model can inform the design of features at various levels of cognitive skill acquisition.

Changing the Modality of Cues

The transition from local control to remote control is inherently complex, not only because the design must represent all of the relevant information required for diagnosis, but also because the performance of skilled operators is likely to depend on a pre-existing repertoire of cues (Coderre, Mandin, Harasym, & Fick, 2003; Konecni & Ebbesen, 1982). As will be discussed, system designs cannot simply provide alerts for outcomes. They must present those alerts in a way that is consistent with practitioners' expectations. These expectations are usually based on the practitioner's prior experiences.

An example of this phenomenon is provided by the development of electronic fly-by-wire controls in the F-16 fighter jet. Although fly-by-wire systems provided a higher degree of control over the aircraft, tactile feedback in the control stick was lost (Endsley, 1996). In an attempt to compensate for this loss of tactile feedback, engineers provided the necessary information on visual displays (Kuipers, Kappers, van Holten, van Bergen, & Oosterveld, 1990). Nonetheless, otherwise skilled pilots had problems determining airspeed and maintaining flight control with fly-by-wire systems.

Cue-based perspectives of expertise provide a valuable insight into why the transition from tactile feedback to visual displays resulted in a deterioration of performance. Brunswik (1955) proposed that, of the infinite number of features available, only some of these features are associated with the target outcome. At the same time, human operators are limited in the number of available features that they can process simultaneously due to the limitations of working memory (Miller, 1956). However, skilled practitioners are able to bypass this limitation through the automation of processing.

According to the recognition-primed model, skilled practitioners use cue-associations in memory to automate the recognition of critical events (Klein, 1989). However, to be acquired, cue-associations require specific circumstances. First, the feature must be salient to the operator. Second, it must frequently predict or co-occur with a single criterion event. Third, the practitioner must have sufficient experience with the feature and event for the two to become strongly associated in memory (Klein, 1989). If these conditions are met, recognition is eventually triggered in the presence of the relevant feature without conscious attention.

A *stick shaker* under manual flight controls meets these conditions. First, the stick shaker is a dramatic and salient form of tactile feedback to indicate an impending aerodynamic stall. Second, it is a reliable indicator of an event that, left unaided, will eventually result in a catastrophic outcome. Therefore, with sufficient experience with the *stick shaker* and concurrent aerodynamic stall, pilots develop an association between the two events in memory. As the pilot gains and maintains this experience, the activation of this association becomes automated and eventually non-conscious. This automation of processing enables the allocation of working memory resources to other tasks (Ericsson & Lehmann, 1996).

When electronic, fly-by-wire control systems were first introduced in the F-16, engineers removed the tactile feedback provided by the stick shaker. Consequently, test pilots could no longer access a cue which they had come to rely upon (Kuipers et al., 1990). They were also expected to develop a new feature–event relationship by using visual instruments in place of tactile feedback. Inevitably, this process degraded performance, to the point that artificial *stick shakers* were introduced so that pilots could capitalise on their existing feature–event relationship.

Although there is little doubt that the pilots would have eventually acquired the new cue-based association in memory, they were likely to experience a loss of performance in the interim. In complex, high risk, high consequence industrial environments, the consequences of changing the nature of feature–event/object relationships can be significant. Therefore, where appropriate, the ideal solution for introducing new technologies involves maintaining or capitalizing on feature–event/object relationships with which the practitioner is already familiar.

Miscueing and Poorly Differentiated Cues

On 14 August 2005, Helios Airways Flight 522 crashed into the side of a mountain northeast of Athens, Greece, killing all 121 passengers and crew. In addition to a significant loss of life, the accident was notable because the subsequent investigation revealed that the pilots had received and acknowledged multiple warnings (Air Accident Investigation and Aviation Safety Board, 2006). However, as this chapter will argue, the mode by which those warnings were delivered prevented the pilots from actioning an appropriate response.

The initial warning presented to the crew of Helios 522 was far from insurmountable. The aircraft took off with the cabin pressurization system set to manual, rather than automatic (Air Accident Investigation and Aviation Safety Board, 2006). As the aircraft climbed, the air pressure in the cabin gradually decreased, resulting in a lack of breathable air. The pilots could have quickly resolved the pressurization issue if they had recognized what was occurring. Through the subsequent investigation, it was evident that the pilots received a number of warnings from the automated alerting systems. In fact, the cabin altitude warning alarm sounded repeatedly from 12,040 feet.

Despite the fact that the pilots received numerous warnings, they appear to have succumbed to a phenomenon known as miscueing. Miscueing refers to the activation of an inappropriate association in memory by a salient feature, thereby delaying or preventing the accurate recognition of an object or event (Rowe, Horswill, Kronvall-Parkinson, Poulter, & Mckenna, 2009). Under certain circumstances, miscueing can result in an expert practitioner's speed and diagnostic accuracy decreasing to that approaching a novice (Shanteau, 1992).

In the case of Helios 522, the altitude warning received by the pilots was similar to the more common take-off configuration warning (Air Accident Investigation and Aviation Safety Board, 2006). Consequently, when the alert sounded, the pilots became fixated on explanations that related to the configuration of the aircraft. This was despite the fact that the configuration warning was only meaningful when the aircraft was on the ground. In fact, the pilots' fixation was so focused on configuration-oriented explanations of the event, that they failed to attend to mounting evidence of a pressurization problem. Indeed, when the ground engineer specifically asked about other indicators of pressurization, the captain stated that the warning lights were off. The flight data recorder indicated otherwise.

The circumstances of this accident highlight one of the risks associated with recognition-primed decision-making – features usually develop associations in memory with the most commonly paired event. This association can become so strong that it potentially overrides the activation of other possible associations and interpretations.

In remote, control environments, the practitioner relies entirely on standardized features. If those features are not sufficiently differentiated, there is no scope to develop more nuanced associations. Under such conditions, the practitioner may never be able to avoid miscueing errors. Importantly, due to their extensive experience and automation of processing, experts have stronger associations in memory than novices (Cellier, Eyrolle, & Marine, 1997), and are potentially more susceptible to miscueing errors.

From a design perspective, miscueing errors can be avoided by differentiating prompts and warnings. This can be achieved in two ways, the first of which is to make sure that each cue is sufficiently distinct. The second is to add secondary cues that help practitioners differentiate possible options activated in memory. Both approaches have advantages and disadvantages depending upon the skills of the practitioners.

Following the release of the Helios 522 air crash investigation findings, the Federal Aviation Administration released a directive to reduce the ambiguity of pressurization alerts. The relevant aircraft models were to be fitted with two additional cockpit-warning lights that differentiated between problems with cockpit pressurization and problems with the take-off configuration.

Designs and Incorrect Relationships

One of the challenges inherent in centralized control environments is that operators are often removed from the components of the system over which they are responsible. Consequently, they do not directly observe the causes of system events nor the outcomes of their responses, thereby reducing the opportunity for the operator to develop a mental model of the interactions between system components.

For the first generation of supervisory control systems, this was not a particularly significant issue because most of the operators had previously worked with the physical system. For example, as late as the 1990s, electrical distribution engineers energized and de-energized high-voltage power lines in the field. This experience of travelling to remote sites provided the engineers with a concrete conceptual model of the interdependencies between the lines. When the power industry transitioned to centralized supervisory control systems, and the engineers were tasked with monitoring lines hundreds of kilometres apart, their experience in the field enabled them to draw on pre-existing associations between network components to recognise the flow-on effects of their actions in the control room. For operators with this level of experience, it wasn't necessary to provide an intuitive representation of the network as part of the control panel.

Problems can begin to emerge when the first generation of supervisory control system operators retire and a new generation of operators takes their place. This is because the new generation of controllers have missed the opportunity to operate in the field and, therefore, have not necessarily developed the associations and mental models born of that experience (Hoc, Amalberti, & Boreham, 1995). Their knowledge of the functions and dependencies of the system will be based, almost entirely, on the design of the supervisory control system interfaces. As a result, the design of these interfaces is likely to have a significant impact on overall system performance.

The implications of an increased dependence on interface design is illustrated by recent research in the context of power control, whereby the supervisory control system was centred on a map of the network that had no real-world basis (Loveday, Wiggins, Harris, Smith, & O'Hare, 2013). Where the first generation of operators had been able to use the interface more or less effectively, presumably by drawing upon pre-existing associations in memory, those operators who had never worked with the lines directly struggled to recognise key dependencies. For example, they failed to recognise that, if one line failed due to a bush fire, other lines in close geographical proximity were also likely to be impacted. Similarly, they would ascribe associations between lines when no direct association existed in reality.

It was readily apparent why next-generation operators were frequently unable to identify when power lines were at risk: the system only clearly communicated functional dependencies between the lines, i.e. where the electricity current would flow if a line failed. The system did not necessarily illustrate the geographic proximity of lines, since this was assumed knowledge. In fact, even when the lines ran immediately alongside one another in reality, they were sometimes represented on the supervisory control system interface as perpendicular lines at opposite sides of the display.

In the case of Loveday, Wiggins, Harris et al. (2013), less experienced controllers' tendency to assume that unrelated lines were in imminent danger probably reflects the application of a robust, pre-existing association from memory. In particular, inexperienced engineers may have drawn on general heuristics associated with the interpretation of maps, namely, that the distance between objects is to scale. Consequently, changes to the design of a system not only required novices to develop new domain-relevant associations in memory, but to override their pre-existing general heuristics.

As this example highlights, it is generally more efficient to design interfaces to capitalise on the pre-existing heuristics of users. This obviates the need to develop and maintain novel cue-based associations, thereby reducing training time and increasing the reliability of responses, particularly in high-risk, time-constrained situations.

Alarm Floods

When designing a system, it is tempting to include the provision to alert users to all of the various system events that occur, particularly during non-normal situations. There is an assumption that this enables accurate and efficient problem resolution, since the operator is made aware of the status of the various components of the system. However, from a cue-based perspective of decision-making, this strategy may actually degrade users' diagnostic performance, since multiple cues will be triggered simultaneously with the associated demand for cognitive processing.

Referred to as 'alarm floods' operators can become inundated with a greater number of alerts and prompts than they are capable of processing (Schleburg, Christiansen, Thornhill, & Fay, 2013). For example, during an earthquake in Christchurch in 2012, control room operators were confronted with approximately 150 prompts as different power lines failed. Although the cause of the failures was immediately obvious to the controllers, they spent a considerable proportion of their time acknowledging prompts that contributed very little to their management of the situation.

This approach to design is at odds with models of cue utilization, since human operators are capable of processing only a few sources of information simultaneously. Where experienced operators manage this limitation by attending only to those features that have the strongest association with critical events (Klein, 1989), inexperienced operators can find the processing or filtering alerts and alarms overwhelming. Consequently, even an expert practitioner will either ignore system prompts so that the situation can be managed or, alternatively, may delay a response to the situation by acknowledging system prompts that, otherwise, have no impact on the situation.

From the perspective of the recognition-primed decision model, whether the user is a novice or an expert, there is little to be gained by alerting operators to every system event, particularly during non-normal situations. Instead, designers should seek to identify the most important information and craft fewer features that offer a more holistic indicator of the cause. This will ensure the activation of appropriate feature–event relationships in memory and thereby improve response latency, particularly in time-constrained situations (Julisch, 2003).

Feature Design Options

The case studies discussed in this chapter demonstrate the utility of the recognition-primed model in the identification of flaws in the design of system interfaces.

The remainder of the chapter will take a more proscriptive approach, discussing potential guidelines for effective feature design. To begin, we can infer several specific guidelines from the case studies discussed:

1. Cue associations are feature dependent – the same information in an unfamiliar modality may not necessarily activate the appropriate cue association.
2. Features must be sufficiently distinct to ensure that an appropriate association is activated.
3. Practitioners may have domain-general cue associations, which should be considered when designing domain-specific features.
4. Practitioners can only process a limited number of features simultaneously.

There are, however, many other issues that must be considered when designing features for interface displays. For example, system designers might consider whether features should be static or dynamic, whether features are more effective when they guide the user towards additional information or when they cue the type of event occurring, whether it is more effective to provide a single, integrated feature or multiple discrete features, and whether it is more effective to present users with iconic (concrete) or abstract features. As the remainder of this chapter will discuss, the recognition-primed framework suggests that the answers to these questions may depend upon the practitioners' level of expertise.

Static or Dynamic Features

When designing features, one of the key decisions that must be made is whether the feature should be static (i.e. either present or absent) or incorporate elements of dynamism (i.e. continuously present but changing). From a recognition-primed perspective, both static and dynamic features have advantages and limitations in diagnostic processing performance.

The differentiator between static and dynamic features is that static features may lack critical information inherent to dynamic features (Fabrikant, 2005), particularly the rate of the change, the magnitude of the change, and the presence of the change itself (Tversky, Morrison, & Betrancourt, 2002). Shanteau (1992) hypothesized that, because expert performance relies on the activation of appropriate cue-associations, the absence of relevant features should result in expert performance decreasing to a level that is little better than novices. Therefore, if experts begin to rely on the additional information provided by dynamic features, they are likely to perform little better than novices when presented with static sources of information in isolation (Shanteau, 1992).

The proposition that expert performance can be dependent on dynamic information is consistent with the observations of Sebanz and Shiffrar (2009) who reported that, although expert sports people were more accurate than novices in predicting opposition movements, this effect was only apparent when dynamic features were employed as stimuli. When static features were used, the experts failed to outperform novices. Correspondingly, many other researchers have struggled to differentiate expert from novice performance using static features as stimuli, particularly in contexts where dynamic features were more predictive of outcomes than static features (Enis, 1995; Norman, Coblentz, Brooks, & Babcook, 1992; Sebanz & Shiffrar, 2009; Shanteau, 1992).

An alternative view proposes that expert performance is less dependent on the presence of dynamic information than novices, because experts are able to infer dynamic information from static stimuli. For example, Lowe (2001) asked participants to predict and draw meteorological markings on a map of Australasia, using a static meteorological map of the Australian mainland. Non-experts produced very simple forecast maps compared to those developed by experts, suggesting that experts were better able to anticipate and interpret dynamic environments from static cues. Consequently, although experts may perform equally well when provided with static or dynamic features, non-experts tend to require dynamic sources of information, particularly when reasoning about dynamic environments (Lowe, 2001).

Due to the contradictory evidence regarding the relationship between dynamism and performance, Loveday, Wiggins, Searle, Festa, and Schell (2013) investigated non-expert and expert diagnosis when presented with both static and dynamic interfaces. Specifically, they measured each participant's ability to acquire and utilise interface cues generally, and then contrasted performance under static and dynamic conditions. The results indicated that practitioners with a superior capacity for cue utilization were able to maintain performance whether they were presented with static or dynamic features. By comparison, practitioners with a lesser capacity for cue utilization showed a deterioration of performance under the static conditions. Consistent with Lowe (2001), experts were able to infer the information provided by dynamism from static displays, whereas the non-experts were dependent on the more salient dynamic features.

On the basis of the results reported by Loveday, Wiggins, Searle et al. (2013), designers might infer that it is appropriate to introduce extensive dynamism to their system. However, it should be noted that this research only contrasted dynamic and static displays. There was no consideration of the impact of different components of dynamism. For example, although *rates of change* may be informative, it can also be relatively subtle compared to other aspects of dynamism such as *step changes*. In fact, any dynamism within a system is likely to attract the attention of users whether or not it is meaningful (Bartram, Ware, & Calvert, 2001, 2003). Consequently, although dynamism may be useful as a means of directing operators' attention to system components, designers should endeavour to ensure that dynamism actually provides the information they are seeking to communicate.

Information-Oriented or Cause-Oriented Indicators

When developing system alerts, designers also need to consider whether it is more effective to direct the attention of operators to the cause of a problem, or direct them towards more detailed information that the operator can then use to diagnose the status of the system. While it might seem obvious to simply advise practitioners of the cause of a system change, there are circumstances where the cause of a change in the system state may actually reflect a symptom of a further, root cause.

The limitation of cause-oriented indicators is that many systems are not yet designed to self-diagnose the root causes of system events. For example, in contemporary motor vehicles, a low oil indicator light does not necessarily establish the cause of the low oil pressure. Consequently, until technology-driven diagnoses of system events improves, the choice for designers is actually between relatively broad causal indicators, or indicators that direct operators to attend to additional sources of information, with the expectation that this additional information will enable the operator to establish the root cause of the problem.

The selection of features to initiate operator intervention has important implications for system diagnosis, since the choice of feature is based on the assumption that there is a corresponding event or object in memory. However, there are individual differences in both the frequency and the range of feature–event/ object associations in memory, so that for experts, a response may be initiated relatively rapidly based on the activation of highly specific, pre-existing cue (O'Hare, Mullen, Wiggins, & Molesworth, 2008). For less experienced operators, the same feature may trigger associations with several possible cases, thereby necessitating an additional search of the information available (Klein, 1989). Therefore, for inexperienced operators, cause-oriented indicators may be preferable since they will obviate the need to acquire additional information to establish an appropriate response (Ericsson & Lehmann, 1996).

The difficulty in establishing appropriate cause-oriented indicators can be solved through assessments of expert performance in a range of non-normal events where diagnostic skills are applied. By presenting indicators that correspond to the information sources acquired by experts in resolving a particular problem, non-experts can be directed to those indicators that are most likely to be associated with an accurate diagnosis and solution.

Integrated or Discrete Features

Earlier in the chapter, the phenomenon of alarm floods was discussed. It was noted that one of the ways in which this issue has been addressed is through the consolidation of related alarms. Although a battery of alarms should be avoided wherever possible, it is also important to avoid the removal of critical information during the consolidation of alarms. Ultimately, the correct balance may still require the operator to monitor a large number of features, indicators and alarms.

One of the challenges with integrating alarms is that doing so may reduce the amount of detailed information available to the operator. This information may be critical to optimizing their performance (Simnett, 1996). At the same time, if there are too many alerts, the operator's working memory may become overloaded, to the point where important features are overlooked (Ericsson & Lehmann, 1996). Fortunately, as a practitioner's expertise in the domain increases, so too does the capacity to combine features into more complex patterns. These patterns of features may cue the recognition of specific events, by matching the pattern to similar cases stored in memory (Klein, 1989). This process, referred to as pattern-recognition, allows practitioners to generate more accurate diagnoses than singular features, which may or may not be associated with more than one outcome.

Pattern recognition is also fundamental to how experts are able to monitor systems with many instruments without succumbing to cognitive overload. By processing instrument features as patterns, rather than individually, the operator is able to simultaneously retain and process more information in working memory. This presumably works in a similar fashion to Miller's (1956) description of 'chunking' phone numbers to facilitate storage in memory, i.e. we store phone numbers as three chunks (0423–610–521) rather than as individual digits (0–4–2–3–6–1–0–5–2–1).

Since less experienced operators lack the associations in memory that are required to engage in effective pattern recognition, they are limited in their capacity to monitor a range of individual features simultaneously. Nonetheless, they must eventually interact with the system to build the necessary repertoire of associations. In monitoring contexts, there are several techniques that might be used to facilitate the transition to expertise. For example, some organizations prepare less experienced operators for monitoring roles by initially exposing them to fairly limited components of the system and gradually increasing the breadth and depth of their responsibilities.

For designers, it may be possible to facilitate the performance of non-experts by designing integrated displays that obviate the requirement to monitor different features in isolation. However, the design must be consistent with the patterns of information acquired by experts, so that the cues both communicate the nature of the problem and direct operators to the relevant features that form the basis of the diagnosis. With sufficient exposure to feature–event associations, the practitioner will eventually develop patterns of features that can be extracted from the fully operational system.

Abstract or Iconic Features

It is clear that a pre-existing association in memory can shape the way in which novice practitioners interpret interfaces. In some cases, this can result in misunderstandings (Tversky & Kahneman, 1973). However, it may be possible to design features to make use of these general associations to communicate the status of the system more effectively to non-experts.

Consistent with the rationale that features can be developed from domain-general associations in memory, Perry, Stevens, Wiggins and Howell (2007) distinguished between abstract and iconic features in the context of auditory alerts.

Abstract features are those that have no inherent meaning. Instead, an association is developed with significant events through repeating pairing. Abstract features include the standardized alarms and icons used in most control room settings. Iconic features, by comparison, are those that are designed to exploit domain-general associations to provide inherent meaning. For example, since people are usually unable to hear others when wearing earmuffs, railways use the image of a train wearing earmuffs to identify 'silent' carriages where noise should be kept to a minimum.

Since they already have a repertoire of domain-specific cue associations in memory, expert practitioners are unlikely to benefit from the implementation of iconic features. However, for non-expert practitioners, even a weak association can be deduced from iconic cues. Further, if the system is well designed, at least some of the features will be clearly associated with the relevant outcome, reducing the number of new associations that the practitioner must develop to reach competency. Finally, by capitalizing on pre-existing associations, iconic features will become sufficiently associated with the relevant outcomes to improve the rate of skill acquisition.

Consistent with the view that iconic features are a more effective means of cueing recognition, Perry et al. (2007) reported that practitioners required significantly fewer training trials to learn feature–event associations when the feature was iconic than when the feature was abstract. Similarly, the iconic cues resulted in consistently superior alert recognition accuracy. Nevertheless, it is important to note that iconic cues were also associated with increased response latency. Consequently, although there is compelling evidence to suggest that iconic cues can potentially accelerate skill acquisition, they may also delay responses once a practitioner has achieved a certain level of ability.

Conclusion

This chapter considered the implications of the recognition-primed model of decision-making in the context of system design. Four case studies were discussed in which the design of features interrupted the effective use of cues and resulted in failures. On the basis of these case studies, four broad guidelines for the implementation of system features were proposed. A number of feature design options were also discussed, together with their possible impact on practitioners at different levels of expertise. The key implication is that cues that are designed for experts may not be suitable for non-experts and vice versa. However, by adopting a range of design-related principles, it may be possible to accelerate the transition towards expertise while safeguarding the system.

Chapter 6

The Social Context of Diagnosis

Tamera Schneider and Joseph Forgas

This chapter discusses the effects of mood on diagnostic reasoning in the context of Naturalistic Decision Making (NDM). Mood has consequences for social cognition and judgements, and influences diagnostic reasoning in organizational contexts because it influences what is accessible in memory and the way in which information is processed. Mood effects are most likely when we have to construct responses to novel situations, which are common in the context of NDM. First, we briefly discuss the NDM context. Next, we present theory on the influence of mood on judgements, and research that demonstrates how mood affects social cognition and judgements. Most of this work has been conducted in laboratory settings that are markedly different from NDM settings. However, by considering the NDM literature through a social cognitive lens, we can begin to discern the role that mood may have in these situations as they unfold, beginning with the process of diagnosis. Lastly, we discuss issues related to the integration of these two literatures and potential areas for application in organizational settings.

The NDM Context

The study of Naturalistic Decision Making (NDM) centres on the use of knowledge and experience in dynamic, high stakes and high stress situations. These situations could include an aircraft captain and crew whose aircraft struck a flock of geese during takeoff, a surgeon confronted with a patient's excessive bleeding or dangerously low blood pressure, a space crew experiencing power failure or a medical emergency, or a fire chief and firefighters contending with a burning building and its inhabitants. These complex, real-world situations involve time pressure, inadequate information (e.g., missing, ambiguous, erroneous), ill-defined goals, poorly defined procedures, dynamic conditions, and often noise and interruptions (Klein, 1998; Orasanu & Connolly, 1993). These settings require the observation and resolution of problems, both of which involve myriad decisions, with the goal of minimizing cost to life and property.

Klein (1989) sought to understand how experts make decisions in NDM settings. He conducted extensive retrospective interviews with fire commanders, neonatal nurses, and military personnel, inquiring about their decision making in exceptional, non-routine cases. These inquiries marked the beginnings of the Recognition-Primed Decision model (RPD; Klein, 1989). This model focuses on pattern recognition as a way of identifying problems in dynamic situations and on evaluating what courses of action might work, rather than striving for optimal solutions. Recognizing patterns of cues facilitates the distinction about what is atypical and potentially problematic. Noticing anomalies affords the opportunity to address these potential problems (Orasanu & Connolly, 1993). In doing so, the decision maker encounters recognition-related byproducts, such as expectancies that prepare the decision maker to anticipate surprises and to interpret relevant cues, which reduces overload and enables attention to be directed towards those features that are most important. The outcome is the articulation of plausible goals that help to set priorities and create an action plan that is likely to be successful.

Recognition-primed decision making is facilitated by the accumulation of knowledge, skills, and abilities acquired over time, and in multiple situations. Experts use their *gut feelings* or intuition to *size up* problems quickly and then simulate a possible course of action to address the problem (Klein, 1998). Decision making is based on evaluating a particular course of action through mental simulation, by imagining how an action will be carried out, and then identifying weaknesses in the plan. Klein (1998) noted that, in NDM settings, experts typically evaluated only a single option, prior to an intervention. Their goal was to identify an option that would be broadly appropriate under the circumstances. Simon (1957) refers to this type of decision making as satisficing. It contrasts with traditional approaches to decision making that posit that decision making is accomplished by considering all of the options in a comparative analysis. Satisficing distinguishes experts from novices, who cannot rely on experience but instead, utilise time-consuming logic to compare different potential courses of action (Klein, 1998).

The critical components of the RPD model are: (1) sizing up the dynamic situation so that a viable course of action is recognized; and (2) evaluating that course of action with mental simulation. The first component focuses on discerning typicality and, on the basis of this process, expectancies, relevant cues, and plausible goals emerge. Typical courses of action become apparent during the evaluation process. When a situation is atypical, it requires diagnosis and further attention. Therefore, accurate and efficient diagnosis is critical for effective decision making. It requires the development of a repertoire of feature–event/object relationships in memory, and engages bottom-up and top-down processes for gathering and interpreting information. Nevertheless, even with sufficient knowledge, information, and appropriate decision-making skills, the nature of NDM situations, and the nature of decision makers themselves can lead to decisions with poor outcomes (Orasanu, 2005).

Aspects of the environment can amplify the stressfulness of NDM situations and can significantly influence the nature of decision making. For example, changes in ambient noise are associated with increases in the frequency of performance errors (Broadbent, 1957) and encourage a narrowing of attention on central cues, often to the neglect of peripheral cues (Baddeley, 1972; Hockey, 1970). Environmental stressors such as noise can also reduce working memory capacity that, in turn, limits the ability to acquire and interpret information (Smith

& Broadbent, 1981; Hamilton, Hockey, & Rejman, 1977). Situations that are unfamiliar and ambiguous require shifts in attention and impose greater demands on working memory for information search, diagnostic strategies, and evaluating courses of action. These situations are most vulnerable to the influence of stressors (Orasanu, 1997) and it is in these situations that decision making is most likely to be influenced by emotional arousal. On the other hand, in situations that are familiar and unambiguous, and where recognition-based responses are available, emotions are likely to have little influence on decisions (Forgas, 1995).

In addition to the influence of environmental features, the physical and mental state of decision makers can also undermine decision-making efforts. People tend to evaluate or interact with stressors in the environment in unique ways with evaluations of ongoing stressors ranging from *challenge* to *threat* (Lazarus & Folkman, 1984; Schneider, 2008). For example, decision makers interpret a task as challenging when a situation is personally relevant and where there is a perception that they will be able to cope or even gain mastery over the task. Conversely, decision makers will interpret a task as threatening where they perceive that the demands of the task exceed their available coping resources.

When the demands of the decision task are perceived predominately as a threat, a range of psychophysiological reactions ensue that can then interfere with the efficiency of the decision-making process (Blascovich & Mendes, 2000; Schneider, 2004). For example, individuals who interpret a task as threatening will also experience a greater level of negative affect (Schneider, 2004), and will perform at a lower level than their counterparts who interpret the same task as a challenge (Blascovich, Seery, Mugridge, Norris, & Weisbuch, 2004; Gildea, Schneider, & Shebilske, 2007; Tomaka, Blascovich, Kelsey, & Leitten, 1993; Schneider, 2004, 2008). Physiological arousal (referred to as *emotion* in more dated literature) had been thought to play a key role in explaining poor performance (Baddeley, 1972; Easterbrook, 1959). However, physiological arousal is evident in both individuals who interpret a task as a challenge and those who interpret a task as a threat. It is their cardiovascular hemodynamic pattern of arousal that differs. Individuals who interpret a task as a challenge will tend to have a greater proportion of their blood supply directed to the periphery, presumably to improve their muscle function (Schneider, 2004, 2008). By contrast, threat appraisals tend to evoke a retention of the blood supply within the central nervous system, a greater level of negative affect, and, ultimately, poorer task performance (Schneider, 2004, 2008; Schneider, Rench, Lyons, & Riffle, 2012). The implication is that individuals who perceive a task as a challenge appear poised to *approach* the situation whereas threatened individuals tend to adopt behaviours that are more *avoidant*.

Emotion and Decision-Making

Classical philosophers paid considerable attention to the role of affect in human affairs. Affect was thought to be irrelevant at best and, at worst, disruptive to thinking and likely to derail rational decision making. Plato considered affect a more primitive, animalistic mode of responding that was incompatible with reason. Emotions were thought to overwhelm and subvert the rational. Contemporary theorists such as Freud also thought that emotions had a dangerous influence on both thoughts and behaviour.

More recent scientific advances offer a different view, with research over the last few decades revealing that affect is not only useful, but it is an essential component of rational behaviour. Early evidence can be drawn from Damasio's inquiry into Phineas Gage, and patients like him, who had experienced damage to affective areas of the brain but not to cognitive areas (Damasio, 1994). Such affectively impaired individuals tend to make disastrous decisions and their lives suffer accordingly. Therefore, emotions are thought to serve an adaptive function, regulating the decision-making process. Recent neuroscience research confirms the view that centres of emotional control, such as the prefrontal cortex, modulate emotional arousal (amygdala responsiveness) (John, Bullock, Zikopoulos, & Barbas, 2013). The implication is that emotion and cognition work in concert to adapt to the demands of stress-related tasks. Our stance is that affect has neither a universally beneficial nor disruptive influence on the decision-making processes. Rather, whether the influence is adaptive or maladaptive depends on the nature of the task or context, the information-processing strategy used, and the characteristics of the decision maker. Consequently, affect can facilitate effective decision making (e.g., gut feelings, intuition), but it can also contribute to cognitive and judgemental errors.

The Affect Infusion Model (AIM) is an integrative theory of research findings concerning the influence of affect on social judgements (Forgas, 1995, 2001). Affect infusion is the process whereby affectively loaded information influences and becomes incorporated into cognitive and judgemental processes, entering into deliberations and influencing decisions. In principle, affect determines mood, which, in turn, influences what and how people respond to information. It is an active filter for incoming information, thereby impacting the content and valence (positivity/negativity) of memories, judgement and, ultimately, behaviour.

AIM states that mood is *less* likely to influence judgements and decision making when people access directly stored responses (e.g., cues), such as in familiar or unambiguous situations, or when the motivation to reach a particular outcome is high (e.g., maintaining a positive mood). Mood is *more* likely to influence decision making when responses need to be created or constructed, such as occurs in unfamiliar or ambiguous situations. There are two main processing strategies associated with responses in generative contexts: the heuristic processing strategy, and the substantive processing strategy.

The heuristic processing strategy is characterized by the use of only a limited range of information and whatever shortcuts or simplifications are readily available (see Forgas, 1995). This type of processing is most likely when the target is simple or highly typical, and when the situation does not demand detailed consideration. The primary mechanism for affect infusion under heuristic processing seems to be the affect-as-information account, where judgements are based on inferences from the prevailing affective state (Clore & Parrott, 1991; Schwarz & Clore, 1988). Judges ask themselves, 'How do I feel about it?' and mistake their current feeling (due to pre-existing mood) as a reaction to, or diagnosis about, the target (Schwarz, 1990).

Substantive processing is characterized by the generation of responses by selecting, learning, and interpreting novel information about a target and relating this information to pre-existing knowledge structures (see Forgas, 1995). This type of processing is more likely when the task is complex or atypical, and when the situational demands require more elaborate processing. It is a strategy adopted in circumstances where simple heuristics, which use pre-existing feature–event/object associations in the form of cues, prove inadequate. Affect priming affords the most parsimonious explanation for mood effects obtained through substantive processing (Bower, 1991). Elaborated in the associative network model (Bower, 1981), the affect-priming principle suggests that affect indirectly informs judgements by facilitating access to affect-related cognitive categories as the affective response spreads activation through associated memory structures. In this case, affective states selectively prime associated thoughts and representations that are more accessible and are, therefore, used when responses are constructed using memory-based information.

The AIM model predicts that different processing strategies play an important role in determining the nature and extent of affective influences on judgements (Forgas, 1995). However, mood states may also influence the regulation of cognitive processes (Bless & Fiedler, 2006). Positive affect signals a hospitable environment and fosters a schema-based, top-down processing style where pre-existing ideas, attitudes, and representations are used to assimilate information about the external world. In this case, processing is more creative, open, broad, and flexible. On the other hand, negative affect signals a potentially hostile environment, where processing is more deliberative and attention is narrowed. It promotes a bottom-up, externally-focused processing strategy where responses accommodate external constraints. Positive affect should confer adaptive advantages when tasks are best handled by relying on cue-based associations and past knowledge, and require rapid responses. On the other hand, negative affect should confer functional advantages when a task requires careful environmental monitoring, and involves new and concrete information.

Both field and laboratory experiments have been used to investigate the influence of mood on judgements and behaviour, and particular insights on the process have emerged through research in the context of eyewitness memory (Forgas, Laham, & Vargas, 2005). For example, participants who were induced into a positive mood during the recall of eyewitness testimony had incorporated more misleading information into their judgements than participants who were induced into a negative mood. Further, those participants with a positive mood expressed relatively greater confidence in the accuracy of their judgements. Similarly, Unkelbach, Forgas, and Denson (2008) noted that positive mood was associated with a greater level of stereotyped judgements concerning Muslim targets, whereas a negative mood was associated with decreased stereotyping. Finally, Forgas and East (2008) asked participants to judge whether a videotaped target was being deceptive, and observed that those participants who were induced into a negative mood were more accurate in their discrimination of deceptive behaviour. In combination, the outcomes of this research suggest that positive affect tends to promote attention to a broader range of features and the activation of cue-based processing, whereas negative affect tends to promote a narrowing of attention and increased rumination.

Mood and Diagnosis in NDM Settings

Since mood can act as a filter for task-related information and thereby selectively activate memories, it has important implications for diagnosis (e.g., hostile versus hospitable environment). At one level, it can facilitate the activation of appropriate cue-based associations, which ensures a rapid response to a situation. At another level, changes in mood can also potentially bias decisions when they are perceived as inconsistent with task-related goals. As a consequence, affect informs decisions by providing an experiential link to evaluate situations that may be confronted.

People sometimes respond to *intuitions* to avoid options that are perceived as *bad*, which might simply mean that the option is perceived as inconsistent with underlying motivation. With time pressure and uncertainty, decision makers rely on past experience or mental models to quickly generate action. Affect can lead to mistakes when mood is permitted to influence the desire for a rapid response or, alternatively, it can capture past experience quickly, thereby leading to effective outcomes. Whether the influence of an affective state is helpful or harmful in naturalistic decision situations will depend largely on a range of, as yet, poorly understood contextual factors. Moods may also influence the use of different types of processing strategies, and the processing style adopted may, in turn, influence the efficacy of decision making in naturalistic situations.

In the operational context, positive affect is likely to result in a greater reliance on pre-existing knowledge and the tendency to detect fewer anomalies in a situation in comparison to a neutral and negative mood. In contrast, negative affect would be expected to result in increased attention to the external environment and the tendency to detect a greater frequency of situational anomalies. Since attention is focused externally, any anomalies in the environments should be detected more rapidly, compared to neutral and positive mood.

NDM settings are characterized by dynamic conditions with poorly defined procedures. These situations contrast with controlled laboratory settings where tasks embody well-defined goals, and achievement is relatively easy to assess (Kahneman & Klein, 2009). In laboratory and field research settings, participants are less likely to be required to invent or modify procedures, and are under less intense cognitive and affective demands than decision makers in naturalistic environments. Therefore, the operational impact of constructs such as mood can be difficult to establish.

A first step is to identify the key constructs that can be discerned from naturalistic situations to establish the relationship between mood and decision-making in these environments. Important NDM constructs, such as *sizing up* and *evaluating* the situation, could be operationalized to begin to investigate the role that affect might play in naturalistic situations. Sizing up situations requires judgements about typicality – feature matching and story building – and pattern recognition (Klein, 1998). These are the same constructs that form the basis of effective diagnosis.

Initial research might focus on identifying the influence of pre-existing affective states on the recognition of feature–event/object relationships, anomaly detection, and the automaticity of judgements. Previous research suggests that positive affect may promote the use of pre-existing heuristics and top-down reactions, whereas negative mood facilitates greater attention to the external situation. Other key constructs might include decision-makers' expertise, and the accuracy and breadth of their knowledge, experience, and prior, pre-determined, feature–event/object relationships in the form of cues. In some cases, greater levels of expertise may moderate the impact of mood, as expert decision makers may be readily able to retrieve accurate cue-based responses to an operational task. However, when a situation is ambiguous, the direct access of stored responses is less likely and, in this case, negative mood may have adaptive consequences, increasing judgemental vigilance and contributing to the more effective and efficient detection of anomalies.

Conclusion

Clearly there are many questions that remain in developing an understanding of the relationship between mood and diagnostic accuracy and efficiency in high risk, high consequence environments. Most importantly, the relationship between mood and cue salience needs to be established with a view towards understanding the moderating impact of expertise. This research needs to be undertaken beyond the laboratory environment to ensure that the complexities and demands of operational contexts are taken into account. If mood does impact decision making in naturalistic settings (e.g., anomaly detection, pattern recognition), and the primary mechanisms can be established, then this information will provide the basis for diagnostic support tools to safeguard systems during the process of skill acquisition.

Chapter 7

Diagnosis and Instructional Systems Design

Mark W. Wiggins

Clearly, one of the goals associated with developing an understanding of diagnosis and the basis of diagnostic skills is the capacity to develop pedagogical initiatives that will enable the acquisition of diagnostic skills across a range of applied settings. Cue-based approaches to skill acquisition are significant in this regard as they offer a potentially effective means of reducing the cognitive load imposed during and following learning, thereby enabling the development of associations between features and events/objects more rapidly than might occur otherwise.

Cue-based learning also potentially reduces the time required for learners to identify key features, or those features with the greatest predictive capacity. In essence, the specification of key features obviates the requirement for multiple trials that are required to 'discover' those features that are more predictive than others in determining the subsequent performance of the system. Finally, the effective identification of key features for the learner potentially reduces the level of anxiety associated with skill acquisition, thereby increasing the cognitive resources available for the learning process.

The Nature of Cue-Based Learning

Cue-based learning takes many forms. However, in the context of diagnosis, the central tenet of cue-based learning is the specification and association between those features and events/objects that yield greatest predictive validity (Reischman & Yarandi, 2002; Wiggins & O'Hare, 2003a). Typically, the process involves directing the learner's attention towards a key feature and then asking the learner to observe the subsequent event/object (Weaver, Newman-Toker, & Rosen, 2012). Over a series of trials, there is an assumption that the learner associates, in memory, the feature with the event/object to form a cue.

Directing the learner's attention to key features can be more or less overt, depending upon the nature of the task and the aims of the instructional system. In the case of the Discovery Method, learners interact with a system on a number of occasions to 'discover' the key associations (see also Chapter 9, this volume). According to Logan and Schneider (2006), the cues, in this case, are transparent, such that the relationship between the constituent elements is not immediately obvious to the learner. The disadvantages with this approach are the time required to undertake multiple trials, together with the dependence on the accuracy and efficiency of the feedback provided. There is significant evidence to suggest that both the quality and the timeliness of feedback impact significantly the rate at which learners acquire cue-based associations (Kontogiannis & Linou, 2001).

The primary advantage associated with the Discovery Method is, arguably, its capacity to strengthen memory-encoding strategies so that they are relatively resistant to memory loss. This is due, in part, to the discrimination that occurs as part of the discovery process between those features that are more or less predictive of specific events (Morrison, Wiggins, & Porter, 2010). The result is a repertoire of features with greater and lesser predictive validity, thereby providing a capacity to both isolate predictive features and actively disregard features that hold little or no predictive validity.

The alternative to the Discovery Method is a progressively more directed process in which either the key features are revealed to the learner or both the features and associated events/objects are identified and the relationship between the two is illustrated. The advantage in employing less transparent cues (cf. Logan & Schneider, 2006) is that the process retains some of the characteristics of the Discovery Method insofar as active involvement is required on the part of the learner to at least identify the associated events/objects. However, consistent with the Discovery Method, the success of the process is dependent upon the quality of the feedback provided (Lee & Vakoch, 1996).

Identifying the most predictive features and events/objects for learners holds a number of advantages, including the efficiency with which cues can be acquired. However, unless some consideration is given to recognizing those features that are less predictive, learners can be placed in situations in which features may emerge, the predictive validity of which will be unknown (Cohen & Freeman, 1996). This may lead to distraction and uncertainty as the learner questions whether the novel feature holds a greater level of predictive validity than the feature that has been identified during training.

An allied issue concerns the extent to which features and events/objects that are not intrinsically acquired by learners can, in fact, provide the basis for internalized cues. For example, Anderson (1993) notes that, in the case of productions, condition and action statements can only be associated when both are resident simultaneously in working memory. This requires active engagement with the environment to establish precisely the nature of the condition and its relationship to the subsequent action.

The nature of cues is similarly dependent upon an internalized process whereby there is a realization, on the part of the learner, of the relationship between the precise value of a feature and an event/object. This realization is often spontaneous and is evident externally through a relatively rapid increase in skilled performance. In essence, it represents the learner's recognition of the delineation between more predictive and less predictive features of the environment.

The rapidity with which cue-based associations occur suggests that they can form in the absence of conscious processing. This lack of conscious deliberation constitutes tacit knowledge, and may explain the difficulty that experts experience in attempts to articulate the specific features on which their behaviour is based, particularly in relation to psychomotor performance (Nisbett & Wilson, 1977; Patterson, Pierce, Bell, & Klein, 2010). However, the nature of these associations also prevents inquiry as to whether feature–event/object relationships have been acquired and/or whether they represent accurate and efficient associations under the circumstances. For example, inefficient or inaccurate associations may only become manifest in specific situations where success is dependent upon the accuracy and/or efficiency of feature–event relationships. This presents difficulties in high-risk environments where the consequences of errors can be especially significant.

From the perspective of Instructional Systems Design (ISD), the difficulty in assessments of feature–event/object associations lies in the requirement to develop formative evaluation strategies that specifically target the feature–event/object associations that constitute the expected learning outcomes. Therefore, there is a need to progress beyond knowledge-based assessments to skill-based assessments that are designed to both examine the application of specific feature–event/object associations and identify those specific associations that require additional reinforcement.

By their very nature, cue-based assessments must be contextualized, since it is the context that will provide the trigger for the application of the feature–event/object association. Moreover, it is important to ensure that the appropriate triggers are in place. Providing inappropriate or inadequate features is unlikely to trigger the application of a cue, unless the learner has developed sufficiently generalized associations that are capable of coping with differences in features.

The results of transfer of training research suggests that training outcomes tend to be particularly narrow and that learners typically have difficulty in translating the information acquired during training to applied practice (Blume, Ford, Baldwin, & Huang, 2010). Part of this difficulty relates to the dissociation between training environments and operational environments. For example, there is strong evidence to the effect that maximizing the transfer of cognitive skills requires the acquisition of skills in the context in which they are expected to be applied (Richards & Compton, 1998).

The association between the acquisition and application of cognitive skills is encapsulated within the notion of situated cognition in which there is an assumption that cognitive skills are contextualized, particularly at the earlier stages of skill acquisition (Clancey, 1997). However, an important distinction needs to be made between cognitive strategies, and the key features that represent the application of the strategy within a particular situation (i.e. the skill). For example, cognitive strategies might be regarded as architectures or frameworks into which are inserted contextualized representations of features and events/ objects. As a consequence, it is not the capacity for target detection that is acquired during skill acquisition, but the detection of a target within a particular context.

Support for the existence of generic cognitive frameworks can be drawn from the extensive literature concerning the application of cognitive heuristics (Shah

& Oppenheimer, 2008). In particular, their pervasiveness amongst decision-makers, even when they can clearly be associated with instances of error, suggests that the mechanisms through which heuristics function do not necessarily rely on conscious processing. The trust that is invested in the accuracy of heuristics possibly reflects the fact that, in most cases, the application of heuristics leads to successful outcomes (Gigerenzer & Goldstein, 1996).

The double-edged nature of heuristic-based processing is evident in the literature concerning the relative utility of heuristics. For example, Kahneman (2003) contends that heuristics are fundamentally error-prone, and that decision-makers need to be taught to overcome the temptation to engage in relatively simplistic reasoning strategies. In contrast, Todd and Gigerenzer (2003) argue that heuristics are generally accurate and that it is only in a relatively narrow range of situations where heuristics will be applied inappropriately and will lead to error. Similarly, cues can be either accurate or inaccurate, depending upon the circumstances in which they are applied.

Approaches to Cue-Based Training

Error management training is an approach to learning that enables the exploration of the contexts within which responses are more or less appropriate (Carter & Beier, 2010). Driven by the learner, it facilitates hypothesis testing and inevitably results in error. From an instructional perspective, these errors are employed as learning opportunities whereby the trainee establishes the boundaries within which appropriate behaviour can be applied (Ivancic & Hesketh, 2000).

In a cue-based framework, it might be argued that error management training is enabling the relationship between features and events/objects to become better refined so that the context for the application of a cue becomes more specific and, therefore, more accurate. The difference between this type of approach to instruction and more directed approaches is the nature of the cognitive involvement of the learner in the process of association. In error management training, the learner learns appropriate associations by avoiding inappropriate cases, whereas the reverse is the case in more directed training environments.

In the case of both error management and more directed approaches to learning, there is an implicit requirement for the learner to be cognitively engaged with the process. However, there is an extensive literature pertaining to observational or vicarious learning and there is little doubt that learning is possible in the absence of the physical engagement with the task.

Although the utility of vicarious learning has been well-established (Jentsch, Bowers, & Salas, 2001), the relationship with cue-based learning has yet to be established. Nevertheless, there is evidence to suggest that learners can develop skills by observing the behaviour of experts (Boschker & Barker, 2002) and by visualizing behaviour (Cooper, Tindall-Ford, Chandler, & Sweller, 2001).

Therefore, it may be the case that the key in developing cues is to ensure that there is a degree of cognitive engagement in the process.

Cognitive engagement is a theme that underscores Cognitive Transformation Theory (CTT), particularly in terms of the acquisition of cognitive skills (Klein & Baxter, 2009). Proponents of CTT contend that the acquisition of cognitive skills involves the categorization of patterns of information that enables the development of accurate and efficient mental models (Wiltshire, Neville, Lauth, Rinkinen, & Ramirez, 2014). Through active involvement in the performance of a task, learners begin to discriminate between nuanced variations in patterns and prototypes, thereby facilitating both greater efficiency and accuracy in perceptual discrimination.

Unlike models of learning that are based on the systematic accumulation of knowledge, CTT involves an iterative process in which feedback involves directing the learner's attention towards key features that will assist in the restructuring of information and the acquisition of a more detailed and/or more accurate mental model. This is a process consistent with cue-directed learning in which signature features are made salient to learners through carefully constructed learning environments.

When to Implement Cue-Based Learning

Knowing when to expose learners to cue-based training is important to ensure the acquisition of appropriate feature–event/object associations. Previous approaches to cue-based learning have met with varying levels of success, and there are a number of different explanations that might be ascribed to the differences in outcomes. One of the most prevalent difficulties for the developers of cue-based approaches to learning is the extent which the features and events/objects that are taught to learners are perceived as valuable or useful in the operational context. The notion of 'face validity' is common to all training regimes. However, in the case of diagnosis, the problem of face validity is especially problematic, since there is the possibility that learners may perceive the extraction of features and events/objects as over-simplifying a task that is, otherwise, regarded as complex. As a consequence, rather than perceiving 'simplicity' as a learning opportunity, it can actually serve to create a lack of trust in the validity of the information being taught.

As occurs in other forms of training, there is also the possibility that feature and events/objects are presented at a stage of learning in which their integration into a broader mental model is not possible. This may be the product of a lack of experience in the domain, or it may reflect the inevitable dissociation that cue-based learning may encourage, potentially at the expense of the development of a detailed mental model. In essence, the latter reflects the notion that, for some learners, the simplification of a complex task is attractive, since it minimizes the demands for cognitive resources, thereby relieving the resources for other, potentially competing tasks. This short-term gain in cognitive capacity belies the lack of a mental model that would ensure the retention of cues in memory and their appropriate and efficient application. The simplistic application of a cue-based association, in the absence of a reasoned understanding of its implications, is unlikely to facilitate effective and efficient performance in the context of dynamic, uncertain situations.

It is this combination of a detailed mental model and a repertoire of cue-based associations that enables successful performance in a dynamic, uncertain operational context (Albrecht & O'Brien, 1993). In the absence of a detailed mental model, the potential exists for less experienced practitioners to be miscued, responding to some features of the situation to the exclusion of other, perhaps equally diagnostic features. These additional elements of the environment may point towards the application of different solutions that may not be immediately evident in the absence of a mental model. Therefore, as a prelude to cue-based training, it is necessary to ensure that the learner has acquired a mental model sufficient to ensure that the cues acquired during cue-based learning are not simply applied routinely, and in the absence of some consideration of the implications of the response.

The acquisition of an accurate mental model is dependent upon a number factors, including the capacity to identify the distinct components that comprise the system, the capacity to establish the relationship between elements that comprise the system, the capacity to retain the model in memory, and the capacity to draw on the model should this be necessary (Schumaker & Czerwinski, 1990). During learning, there are a number of iterations of a mental model, beginning with a rudimentary structure comprising a relatively small number of elements. With experience, the complexity of the mental model increases as new elements are integrated into the existing framework. Finally, it might become evident that a particular mental model is too general for all circumstances and needs to be modified to better respond to the nuances associated with different situations. Therefore, while some mental models might be regarded as particularly robust, others may demand extensive modification.

One of the difficulties for operators lies in the requirement for the modification of a mental model where the validity of the model has been strengthened over a period of time (Jones & Endsley, 2000; Klein & Baxter, 2009). In some cases, challenges to a strong mental model will constitute a revision to a 'world view' and can take both time to achieve and will represent something of a watershed experience for the operator. Part of the difficulty lies in operators recognizing that some of the assumptions on which they were working may be invalid. However, there is also a potential concern for the operator to the effect that other mental models may be inaccurate or inappropriate and therefore may, in themselves, need to be changed. The uncertainty associated with challenges to mental models can create a level of anxiety that impacts performance over and above the difficulties associated with the mental model.

Establishing accurate and reliable mental models is clearly optimal, since the mental model provides the basis on which information emerging from the environment is acquired and interpreted. As a guide for information acquisition, the mental model enables the identification of those features that are most diagnostic of the situation, thereby restricting the search for meaningful information to a relatively limited set that can be processed efficiently in working memory (Kieras & Bovair, 1984). It also enables the rapid interpretation of otherwise complex scenes, since the information available can be interpreted collectively and matched against patterns of information that are embodied within mental models (Rouse & Morris, 1986). This process of pattern matching ensures that scenes can be interpreted quickly, despite their complexity.

In building a mental model, it is necessary to first identify those features and events/objects that are likely to contribute to the successful interpretation of a situation (Lowe & Boucheix, 2011). Associations are drawn between features in terms of the sequence in which they occur and whether different features are associated with identical outcomes. The latter enables the gradual but systematic construction of a more sophisticated mental model in which features that embody similar characteristics coalesce, thereby generating efficiencies in information processing by triggering fewer, but potentially more predictive, associations. In essence, the compilation of associations simplifies what would otherwise comprise a complex set of relationships. With an increasing frequency of interactions, associations can be further refined and a subset of mental models may emerge as subsets of the originating model.

The segmentation of mental models to form new subsets enables the organization of information necessary for retrieval. For example, in playing a tennis match, a less experienced player may activate a generic mental model concerning the rules and etiquette associated with the game. This generic mental model is likely to be cumbersome insofar as it draws a great deal of superfluous information into working memory as a search is mounted to match information to the environment and thereby develop expectations concerning behaviour. The outcome is a significant increase in cognitive load with, potentially, a minimal return in the capacity to accurately anticipate events.

The value afforded by more targeted mental models is evident in experienced practitioners' acquisition of information from those features of a visual or auditory scene that embody the greatest predictive validity (Craig, Klein, Griswold, Gaitonde, McGill, & Halldorsson, 2012). For example, amongst experienced tennis players, visual fixations tend to be directed towards the position of the opponent's tennis racket in preparation to hit the ball (Shim, Carlton, Chow, & Chae, 2005). However, the application of a more generic mental model might direct attention towards the eyes of an opponent, since the inexperienced tennis player has yet to appreciate that eye gaze is not necessarily a reliable predictor of the direction of the return of the ball.

Developing a Learning Environment for Cue-Based Associations

For learners, the initial identification of features for the purposes of feature–event/object associations is a cognitively demanding process. It begins with a capacity to establish, from a range of stimuli, those elements of the visual or auditory scene that represent features that might form the basis of cues. Having identified these features, associated events/objects need to be identified and the validity of the feature–event/object association established. Since this is a process that is generally occurring at the early stages of skill acquisition, learners are also engaged in building and refining mental models. This requires some degree of cognitive involvement in the performance of a task, together with the capability and willingness to compare the desired and actual outcomes with a view towards addressing potential flaws in existing mental models. It is a process that also imposes significant demands on cognitive load since the learner is not only performing the task, but is also engaged in a process of reflection, a feature that is normally characteristic of expertise (Van Gog, Kester, & Paas, 2010).

Given the information processing demands on learners during skill acquisition, the process of feature identification and cue acquisition demands a level of motivation to ensure that sufficient cognitive resources are invested to identify potential features and then establish their association with events/objects. However, cue acquisition also requires the intrinsic capability to extract features from complex environments where there may be a number of features that each might compete for attentional resources.

Arguably, the capacity to extract features from an environment is distinct from the capacity to draw associations between features and events/objects. The latter requires a capacity to draw causal relationships so that the appearance of a feature can be used to infer an event or the appearance of an object. For example, the sound of a train that is increasing in loudness might be used to infer that a train is travelling in the direction of a listener. For a traveller, the impending arrival of a train (event) might signal the need to entrain. However, for a track worker, the sound of a train might be associated with the possibility of injury or death (object). It is same feature–event/object relation but with very different causal references.

The derivation of meaning on the basis of feature–event/object relationships requires the integration of cues as part of a larger mental model in which multiple causal relationships are evident. The prospective nature of this process is similar in function to the prospective dimension of situational awareness in which experienced practitioners demonstrate a capacity to anticipate and respond to threats in the absence of conscious reasoning processes (Endsley & Bolstad, 1994). However, it also implies that the application of cues as part of a reasoning process involves more than simply the application of cues in isolation. Rather, multiple cues are likely to be involved in the interpretation of complex scenes and, depending upon the combination of cues that are triggered, may infer different consequences. For example, road construction warning signs (feature) that appear coincident with road construction vehicles (feature) or speed warning signs (feature), might be associated with hazardous road conditions (event) and, therefore, infer that a reduction in driving speed is necessary (consequence). However, road construction warning signs (feature), in the absence of any other features, may infer that a reduction in speed is unnecessary.

In the context of diagnosis, cue-based associations that provide the capacity to anticipate future events are not simply a product of the association between features and events but reflect a causal association that infers a conditional relationship. It is the conditional nature of this relationship that enables meaning to be drawn, since the association is applicable in some contexts but not in others.

As a cognitive construct, 'anticipation' has been difficult to define. While there is an understanding of anticipation as something akin to expectation, some conceptualizations of the construct imply a level of insight in which operators are capable of anticipation beyond the information immediately available (Van Der Hulst, Meijman, & Rothengatter, 1999). For example, some researchers discuss episodes where soldiers were able to anticipate and thereby pre-empt an ambush (Klein & Hoffman, 1993). The phrase 'something wasn't right' is often heard in these situations and it often seems to mean the difference between victory and defeat.

Arguably, the notion of intuition or insight in the context of anticipation is nothing more or less than the activation of a cue, albeit involving a feature that may not be especially obvious (Klein, Snowden, & Pin, 2011). When probed concerning their response to an event that, apparently, was difficult to anticipate, operators will refer to nuanced features such as the brief glance of an opponent or a slight deviation from a trajectory (Klein & Hoffman, 1993). This change in ambient 'noise' represented the primary feature on which the response was initiated. While it may have been non-conscious at the time, it was, nevertheless, a response based on a feature–event/object relation or cue.

Such non-conscious relationships between features are well-established in Classical or Pavlovian conditioning. In the case of Pavlovian conditioning, associations emerge between so-called unconditioned stimuli and unconditioned responses (Bitterman, 2006). For example, pairing particular foods with particular events (such as salivation) over a period of time results in association whereby the presentation of the food itself is sufficient to trigger the response. As a result, the response is conditioned to occur on the administration of the food. This effect has generally been demonstrated to occur with reflexive behaviours in the absence of conscious deliberation. This effect might explain the non-conscious response to seemingly innocuous features in the environment; an association does exist between features and events/objects but the association is not immediately apparent to the operator.

The application of non-conscious associations may explain cases where operators experience a physiological sense of unease that is not able to be explained. It might be argued that this physiological response reflects an association that is being activated at a neural level. Therefore, there are features in the environment that are not necessarily explicitly and consciously associated with events or objects. However, associations are acquired implicitly through interactions with the environment. In other words, the non-conscious associations that are evident amongst experts may not necessarily have begun as conscious associations. The identification of features, their association with events/objects, and their application, may have all been non-conscious. Therefore, rather than experts forgetting the associations that characterize their performance, these associations may never have been conscious.

The fact that the development of some cue-based associations may be non-conscious has important implications for training since directed interventions may not facilitate the acquisition of these types of relationship. An example involves learning to ride a bicycle where the explicit articulation of features and events cannot substitute for the experience of riding, albeit with training wheels. The implicit nature of the cues necessary for balance means that the feature–event relationships that clearly must be at work, are difficult to articulate. The difficulty in articulating feature–event relationships appears particularly evident in relation to psychomotor cues or those cues that relate to the cognitive notion of judgement (Kleiner & Drury, 1998).

Accurate and efficient judgement requires the application of rapid responses, often in particularly dynamic situations. Judgement is characterized as non-conscious, precisely as a result of the requirement for responses in the face of shifting demands (Kleinmuntz, 1990). One of the difficulties faced by diagnositicians is the temptation to rely on judgement to the exclusion of a detailed, reasoned approach to information acquisition, deliberation, and response (Croskerry, 2009b). Notwithstanding the fact that judgement can be quite accurate, non-conscious reactions to stimuli can result in misdiagnosis, since operators will often respond to parts of a stimulus. For example, in the case of people experiencing phobias, there is evidence to indicate that phobic responses can be initiated with exposure to partial impressions of stimuli (Elsesser, Heuschen, Pundt, & Sartory, 2006). This suggests that the threat-avoidance response is not simply a reaction to the threat itself, but that the feature–object relationship is capable of activation with relatively little feature-related information available. In situations where time is limited, there is an inevitable temptation to respond to a situation immediately that the feature is recognized. More importantly, the response of phobics implies that there is a degree of non-conscious priming at work to the effect that the reaction to a stimulus is incapable of conscious control.

Diagnostic Cues and Skill Acquisition

There are a number of important principles that provide the foundation for the development of skilled performance in applied industrial environments. At a very basic level of skill acquisition, operators need to develop an understanding of the environment within which they are operating. Part of this process involves learners becoming aware of the experiences and behaviour of others in the domain, and using this information as the basis for their own behaviour. This type of behavioural model has been shown to occur following concurrent observations of behaviour in situ, and where the behaviour is recorded (Snyder, Vandromme, Tyra, Porterfield Jr, Clements, & Hawn, 2011). For example, learners have been shown to both recall and apply specific written cases to which they have been exposed. Presumably, this allows the learner to interact with the environment at some level.

However, the utility of cases appears less evident amongst experts within a particular domain. While 'flash-bulb' memories are evident amongst experts (Curci & Luminet, 2009), these tend to occur in extreme situations involving high levels of risk and time-constraint (O'Hare, Mullen, Wiggins, & Molesworth, 2008), and it is not clear whether these memories direct behaviour, or are simply associated with the application of a behaviour. In other words, flash-bulb memories may constitute a level of redundancy in the event that no response is initiated in response to a threat. It reminds the operator of the seriousness of the situation and possibly, of responses that have been initiated, irrespective of whether these responses were successful.

One of the implications associated with our understanding of modelling and case-based reasoning is that the approach tends to be most effective amongst advanced beginners, rather than competent or expert professionals (Williams, 1992). Advanced beginners have sufficient an understanding of the domain to enable them to perceive the environments within which the case or exemplar can be applied successfully. This process is dependent upon the capacity of the advanced beginner to categorize the case during encoding and the precision with which this process occurs. Where cases are miscoded or are encoded in too general a category, there is a risk that the case will not be triggered or will be triggered in an inappropriate context.

Memory encoding is clearly a critical feature for successful case retrieval. However, it can be difficult for less experienced practitioners to retrieve appropriate cases from memory, despite significant efforts on the part of trainers to ensure the association between cases and subsequent contexts within which the cases might be applied (Molesworth, Wiggins, & O'Hare, 2006). O'Hare et al. (2008) suggest that the difficulty lies in the organization of cases during the encoding process. By clearly demonstrating the association between cases and subsequent outcomes, improvements in the transfer of case-based reasoning have been demonstrated. Indeed, explicit statements of the relationship between learned information and the situations and circumstances within which this learned information can be applied have been demonstrated consistently as the most effective means of ensuring the transfer of information to novel contexts.

Given the utility of case-based learning amongst less experienced operators and the fact that the transfer of learning amongst this cohort is relatively constrained, it might be argued that case-based learning facilitates a relatively limited degree of generalization, possibly only to a relatively limited selection of situations. By contrast, the cue-based reasoning that emerges as the learner progresses beyond competence enables the application of preceding knowledge and skills to a broader range of situations. Therefore, the relatively broader level of generalization evident amongst more experienced operators is a product of the qualitative differences in reasoning that are evident amongst operators.

The limitations of generalization that occur amongst competent practitioners may provide the trigger for the development of new reasoning structures that were hitherto unable to be applied. The desire to improve performance, as evident through conscientiousness, is the driver for more efficient and more effective reasoning strategies than might have been available previously. This results in the extraction of features with diagnostic capacity and the subsequent association with events to form cues.

Conclusion

Although the acquisition of cues will eventually occur through sufficient exposure to a combination of features and events/objects, the process of association can be facilitated through structured learning strategies in which the relationship is made evident. However, simply presenting features and events/objects simultaneously is not necessarily sufficient to enable the acquisition of cues in all cases. There needs to be some consideration of level of the skill acquisition at which cue-based learning is applied, since the provision of cues in the absence of a coherent mental model is unlikely to yield improvements in performance.

Cue acquisition is most appropriate during the transition from competence to expertise, where the learner has acquired a repertoire of cases, the features and events/objects from which can be extracted to form cues. Building on the mental model that was established initially in the transition from novice to competence provides a stronger framework for the categorization of cues and, thereby, their activation.

Building learning opportunities for the acquisition of cues requires the articulation of the features and events/objects that comprise cues, and the development of training initiatives whereby the associations can be both formed and then practised. As for all learning tasks, the provision of appropriate and timely feedback ensures the acquisition of cues as quickly and as rapidly as possible. Similarly, it is essential to ensure that the composition of features is a realistic representation of the environment and that, once acquired, they can be embedded within the operational milieu. This ensures that learners have some understanding of the complexity of the environments within which cues are likely to be applied and can attend to the signature features as they emerge.

Cue-based learning should not be considered a panacea for skill acquisition. It is an approach that should be applied judiciously, taking into account the nature of skills, the stage of skill acquisition at which the initiative might be applied, and the nature of the instructional system available. It offers an opportunity for targeted training, particularly where learners have reached a learning plateau or where performance in the operational context has revealed potential flaws or inefficiencies in cognitive processing.

Chapter 8

Diagnostic Cues in Medicine

David Schell and Marino Festa

I believe that we do not know anything for certain, but everything probably.

> Christiaan Huygens. Letter to Pierre Perrault, 'Sur la préface de
> M. Perrault de son traité del'Origine des fontaines' [1763]

Diagnosis and Clinical Decisions in Healthcare

The backbone of clinical reasoning and diagnosis for novice and inexperienced practitioners has traditionally been the hypothetico-deductive model, similar to that proposed and utilized by the Dutch physicist Christiaan Huygens (1629–95). During medical training and in the first few years following graduation, thinking behaviour surrounding an interaction with a patient follows a well-trodden path, familiar to all medical practitioners. Firstly, the presenting complaint is elucidated, followed by the history of the presenting illness, including a history of previous illnesses, which may also include a family history. For the paediatric patient, the history usually comes from a caregiver, and may also include maternal illness and behaviours during pregnancy. A comprehensive clinical examination is then undertaken. Even at this stage, important clinical signs will be sought to increase or decrease the likelihood of any diagnosis already suspected from the history.

At the end of this initial process, the physician will often have a list of provisional diagnoses (or even a single working diagnosis) that will result in a series of decisions that may result in immediate treatment, or a list of investigations that are designed to either support or, more often than not, disprove a differential diagnosis. In reality, few, if any, tests are capable of proving or disproving a diagnosis absolutely. In fact, the likelihood or probability of a diagnosis is simply increased or decreased by a test result. Bayes's Theorem describes how the probability will never reach exactly zero or 100 per cent (no absolute certainty in either direction), but it can still get very close to either extreme (Gill, Sabin, & Schmid, 2005).

Despite the difficulties associated with diagnosing medical conditions with certainty, several authors have highlighted the potential benefit of this systematic approach in terms of diagnostic accuracy and patient safety. Croskerry (2009a), in particular, has described a dual process theory of medical decision making, whereby a linear, analytical, deliberate approach to information gathering, termed System 2 processing, is utilized in situations where the diagnosis is not immediately recognized. In fact, in the case of less experienced practitioners, even when the diagnosis is thought to be recognized, the traditional 'clerking-in' process requires significant additional corroborative information and steers the practitioner towards the System 2 processing model.

This analytical, rule-based approach to diagnosis and decision-making is recognized as effortful and time-consuming. It has the potential disadvantage of delaying appropriate treatment for time-critical clinical situations due to the need for a high degree of confirmatory evidence accessed in a sequential fashion, before sufficient certainty is reached concerning the diagnosis and management plan. This makes it potentially unsuitable for many time-pressed, real-life clinical situations that are likely to be encountered in daily practice where there tends to be a reliance on the application of tacit knowledge.

The acquisition of tacit knowledge is increasingly important once the clinician enters the workplace and applies existing knowledge and skills. In nursing, Herbig, Büssing and Ewert (2001) demonstrated differences in tacit but not explicit knowledge in nurses who successfully completed a critical nursing task. In the past, the traditional physician-based hierarchy and demonstration of clinical skills promoted an apprenticeship model of 'on the job' skill acquisition through informal learning and the modelling of effective behaviours. Even in the contemporary environment, the concept of the 'doctor' as a 'teacher' remains strongly embedded in the profession.

A high level of context-specific knowledge in healthcare has been achieved by organizing practitioners into teams based on specialty and by connecting teams within a specialty through umbrella professional organizations, or colleges for formal and informal networking. Experience within specific healthcare domains allows an increasing number of similar patterns of presentation to be recognized by clinicians. The observation has been made that accurate and efficient diagnostic performance only becomes stable with a high degree of specialist expertise within medicine (Patel & Groen, 1991), and that extensive training and clinical experience is provided to practitioners to facilitate the acquisition of expert knowledge and skills (Ericsson & Lehmann, 1996). Consistent with this view, improved adherence to evidence-based care and patient outcomes tends to occur where domain-specific diagnoses have been managed by an appropriate specialist with specific expertise, rather than by a general physician with a wider base of clinical practice (Einbinder, Klein, & Safran, 1997; Bucknall, Moran, Robertson, & Stevenson, 1988).

Despite the utility of specialization as a foundation for the acquisition of tacit knowledge, the advent of new information technologies are potentially impacting both the capacity and maintenance of tacit knowledge over time. Novice practitioners, in particular, have been faced with the challenge of incorporating new systems of information recording and sharing into traditional processes of care, often without the benefit of a foundational base of tacit information. These technologies have the potential to significantly enhance a clinical situation or encounter by providing rapid, up-to-date and high quality information. Web-based social media, including weblogs, instant messaging platforms, video chat, and social networks are being integrated into the healthcare industry and will necessitate subtle reengineering in the way the way that doctors acquire, integrate and make sense of diagnostic information (Hawn, 2009). In effect, these technologies have added sources of information that were not otherwise available, and have necessitated the development of innovative practices by contemporary clinicians. Medical educators are now faced with decisions concerning the most effective and efficient methods to embed and utilise computer-based decision support, either in the form of diagnostic checklists (e.g. Isabel and Dxplain) or through the use of more advanced artificial intelligence systems (Bond, Schwartz, Weaver, Levick, Giuliano, & Graber, 2012; Ramnarayan, Tomlinson, Rao, Coren, Winrow, & Britto, 2003).

Diagnostic Challenges in the Hospital Environment

Despite rapid advances in medical technology and improved patient outcomes over the past 50 years, the basic aspects of hospital organization remain largely unchanged, with many vertical silos including those that are physical in nature (wards or departments), and any number of virtual silos that may be structured around professional or educational affiliations, often in a hierarchical manner (Nosrati, Clay-Williams, Cunningham, Hillman, & Braithwaite, 2013). An important vulnerability of the system is the intersection between silos, where failures of communication are likely to occur (Kohn, Corrigan, & Donaldson, 2000). Specific communication vulnerabilities include those between different teams, between medical practitioners and nurses, and at patient handover/ handoff (Alvarez & Coiera, 2006; Cosby & Croskerry, 2004). The critical inter-relationship between the clinician and the healthcare environment has been recognized only relatively recently, particularly in terms of the coordination of diverse disciplines and specialties, bound by organizational rules. Understanding the implications for information acquisition and problem-solving in dynamic clinical situations is leading to a deeper, real-world, understanding of medical diagnosis (Laxmisan et al., 2007). To address these issues, communication tools and checklists have been introduced in many healthcare systems across the world (Ely, Graber, & Croskerry, 2011; Salzwedel, Bartz, Kühnelt, Appel, Haupt, Maisch, & Schmidt, 2013; Thompson, Collett, Langbart, Purcell, Boyd, Yuminaga, Ossolinski, Susanto, & McCormack, 2011). Yet, the most effective use of these tools within existing practices remains to be realized in many cases (Marshall, 2013). Interdisciplinary training programmes, based on real-world processes of care, and sometimes involving simulation, have been adopted to improve patient outcomes and have themselves been able to inform and streamline the safest, most efficient processes of care (Crofts et al., 2008; Marshall, 2013; Hales, Terblanche, Fowler, & Sibbald, 2008). Similar training at an undergraduate level has been used to shape attitudes and inform inter-professional relationships prior to entry into the workforce (Siassakos et al., 2009). Therefore, the concept of professional socialization, which involves the novice practitioner acquiring the habits, beliefs, and accumulated knowledge of a specialist or consultant in his or her chosen field of specialty through education and training, appears to remain at the heart of postgraduate developments of expertise in health.

In an attempt to improve health-related outcomes, many healthcare organizations are emulating the same levels of performance and safety that are being achieved in other high-risk industries, including military aviation, air traffic control, and nuclear power (Carroll & Rudolph, 2006). A number of key features that characterise high reliability organizations may also be recognized in the organizational aspects of healthcare (Hines, Luna, Lofthus, Marquardt, & Stelmokas, 2008). Like other high-risk industries, healthcare utilises highly complex systems that must be coordinated to achieve safety for staff and patients. Multiple decision makers must work together in a complex communication network, and the accuracy and the efficiency of communication between decision makers must be effective. Tightly coupled team members must perform specific tasks that are necessary for success and good patient outcomes. In the case of crises, leadership is deferred to those with the most knowledge of the event, regardless of their organizational rank.

Unlike other high consequence organizations, differentiating inevitable deaths (due to an underlying terminal disease) from those caused by error can be difficult in healthcare, and as a consequence, organizational accountability may be blurred. Until very recently, health in general has been characterized by a 'shame and blame' mentality, where an individual (rather than the system) has been accorded responsibility for an adverse event, including diagnostic error (Reason, 2000). This has potentially inhibited the reporting of adverse events and near misses, the detailed analysis of these events, and feedback to staff with subsequent system improvements (Reason, 2000).

Based on autopsy studies, the rate of diagnostic error in hospital medicine is estimated to be somewhere between 10 and 15 per cent (Graber, Franklin, & Gordon, 2005). In a retrospective case review, Graber et al. (2005) identified 100 patients with a diagnostic error and determined whether the error was cognitive or system-related. Cognition was identified as a factor in 74 per cent of the cases, while system-related factors (including problems with communication, teamwork, processes of care, policies or procedures) were identified as a factor in 65 per cent of cases. Prior to this research, errors in diagnosis and decision-making were usually considered purely cognitive in origin. However, Graber et al. (2005) have illustrated the interplay between cognition and system-related factors, with the latter contributing to missed, delayed or erroneous diagnoses including inappropriate test requests, a failure or delay in receiving results, and/or the incorrect application of results to patient care (Thammasitboon, Thammasitboon, & Singhal, 2013). Therefore, system-related factors, based often on traditional organizational aspects of hospital care (e.g. workload, working hours, physical environment) are increasingly recognized as being important contributors to errors in cognition in healthcare.

Contemporary healthcare, particularly operating theatres and intensive care, are costly and technology-rich environments. Donchin and Seagull (2002) have described the intensive care unit as an 'ergonomic disaster'. It is noisy, lighting is often inadequate, there are multiple distractions, including competing tasks and interruptive conversations, and space may be insufficient. A high level of skill is necessary to interface with the multiple devices that monitor and support a patient safely. Additional complexity occurs due to inadequate standardization involving both the nomenclature used by different manufacturers and the characteristics of user interfaces (the human–machine interface) (Donchin, Gopher, Olin, Badihi, Biesky, Sprung, Pizov, & Cotev, 1995). The volume of data available is enormous and little consideration has been given to the limits of human cognition (Donchin & Seagull, 2002). In an attempt to assist clinicians, each device embodies one or more auditory and visual alarms, sometimes for multiple patient parameters, but also for device malfunction or failure. The purpose of these alarms is to warn staff of a potential change in clinical status, particularly when the clinician is distracted or attention is directed elsewhere. However, given the number of alarms available, the source of any single alarm may be difficult to identify, thereby imposing additional cognitive load, and potentially delaying an intervention (Kaye & Crowley, 2000). Up to 400 alarms per patient per day can occur in some intensive care scenarios, causing difficulty, particularly in distinguishing false alarms from those that are physiologically important (Joint Commission on Accreditation of Healthcare Organizations, 2011). The sheer number of alarms can induce 'alarm fatigue', a state whereby nursing staff, in particular, become desensitized to the multitude of noises and fail to react to an alarm that was initially intended to improve patient safety (Mitka, 2013).

In some areas of the hospital, workload demands can often change rapidly, and events may be unpredictable, such as occurs in the emergency department and the intensive care unit. Clinicians may be responsible for a number of patients simultaneously, each with varying degrees of illness severity and complexity. Croskerry (2002) has likened the process to plate-spinning on sticks insofar as all of the plates must be maintained in motion, without letting any slow and fall. More importantly perhaps is that, as each plate is removed, it is replaced by another.

The optimal number of patients that any individual clinician can manage simultaneously is likely to vary considerably. However, in relation to nursing workload and patient outcomes, the evidence suggests that areas staffed with higher numbers of registered nurses have improved outcomes for patients (Carayon & Gurses, 2005). The traditional role of the senior doctor has been to provide oversight and guidance in patient management to the primary caregivers, the nursing staff and the intensive care team. Tibby, Correa-West, Durward, Ferguson and Murdoch (2004) demonstrated that increased adverse events in intensive care were related to patient factors such as average patient dependency, the number of occupied beds and the presence of multiple, simultaneous management-related issues that compromised the supervisory ability of the nurse in charge. Factors associated with decreased adverse events included the presence of a senior nurse in charge, a high proportion of the shift filled by rostered, permanent nursing staff, the number of admissions and discharges and, surprisingly, the presence of new junior doctors (Tibby et al., 2004). Professional bodies recommend a limit in the number of critically ill patients whose care can be safely managed by a single medical consultant practitioner. However, little data exist to support a threshold beyond which cognitive load is likely to lead to increased error and worse patient outcomes. Nevertheless, a recent meta-analysis supports the proposition that a higher ratio of consultant doctors to patients tends to be associated with reductions in mortality (Wilcox, Chong, Niven, Rubenfeld, Rowan, Wunsch, & Fan, 2013).

Long working hours have traditionally been seen as an important part of residency training for junior doctors, with the long hours and additional patient exposure thought to contribute to the acquisition of expertise. Similarly, senior consultants are often 'on call' for days at a time. As a consequence, there is a significant risk for both acute and chronic sleep deprivation for medical practitioners of all grades. There are a number of well described neurocognitive consequences of sleep deprivation, including reductions in higher-level cognitive abilities, cognitive and psychomotor impairment, lapses in attention, and a reduction in working memory (Gohar, Adams, Gertner, Sackett-Lundeen, Heitz, Engle, Haus, & Bijwadia, 2009; Goel, Rao, Durmer, & Dinges, 2009). Sleep deprivation results in psychomotor impairments of a similar nature to those induced by moderate alcohol consumption, and is associated with an increased risk of human error-related accidents (Goel et al., 2009).

Importantly, there are differences in the cognitive vulnerability of individuals to sleep deprivation, and this may involve alterations in brain function that can be demonstrated using imaging techniques such as Magnetic Resonance Imaging (MRI) or Positron Emission Tomography (PET), and is likely to have a genetic basis (Czeisler, 2009; Goel et al., 2009). Reduced working hours, particularly for junior doctors, have been introduced in many parts of the world. However, reduced working hours have raised other challenges, particularly those relating to the continuity of care whereby the junior doctor is less likely to witness directly the outcome of a series of decisions that were made during the diagnosis and management of a patient. Other important workplace practices have also been associated with reduced junior doctor hours, including the need for more frequent handovers, and the allocation of traditional physician roles to other staff groups, such as the nurse practitioner. In response, there is a growing call to increase the duration of medical training to allow sufficient time for physicians to acquire the breadth of expertise necessary to function safely across a range of medical situations (Kevat, Cameron, Davies, Landrigan, & Rajaratnam, 2014).

Expert versus Competent, Non-Expert Clinicians

The expertise of medical practitioners has typically been connoted based on cumulative experience in the domain. However, even amongst highly experienced clinicians, there may actually be two levels of operators (Loveday, Wiggins, Searle, Festa, & Schell, 2013). These levels are presumed to reflect 'competent non-experts', who rely on prior cases and rules (Rasmussen, 1983), and 'genuine experts', who utilise reliable and efficient cognitive shortcuts (Wiggins, 2006). The assertion that experts utilise cognitive shortcuts is consistent with studies in other industries and domains, which have reported that genuine experts, identified on the basis of diagnostic accuracy rather than experience, are more likely to perform diagnoses using pattern recognition (Coderre, Mandin, Harasym, & Fick, 2003; Groves, O'Rourke, & Alexander, 2003; Norman, Young, & Brooks, 2007). The observation of expert pattern recognition suggests that expert practitioners possess highly refined and strong feature–outcome associations in memory (Coderre et al., 2003). These 'cue' associations represent an association in memory between the features of the patient and a subsequent outcome or illness (Schmidt & Boshuizen, 1993), and are thought to lead to a non-conscious recognition of illnesses or diagnosis.

Cue-based pattern recognition reduces cognitive load during information acquisition, without necessarily sacrificing the depth of processing (Sweller, 1988) and, in so doing, it allows experts to generate rapid and appropriate responses to environmental stimuli (Wiggins & O'Hare, 2003b). For example, in a 'think-aloud' study of gastroenterologists, pattern recognition was evident during seemingly automatic diagnoses that resulted in accurate treatment responses (Coderre et al., 2003). Similarly, Loveday et al. (2013) were able to distinguish competent practitioners from genuine experts by measuring the component skills utilized in rapid pattern recognition (Loveday et al., 2013). In this study, a battery of cue-based tasks was developed within the software package, EXPERTise 1.0 (Wiggins, Harris, Loveday, & O'Hare, 2010) to measure performance on four diagnostic tasks in which the selection and extraction of appropriate cues is presumed to be advantageous. The results indicated that measures of experience, both general and specific, yielded only weak to moderate relationships with performance on the EXPERTise tasks. More interestingly, two distinct groups of participants, formed based on performance across the EXPERTise tasks, differed in the accuracy of their diagnoses during simulated scenarios. Greater accuracy was associated with the faster recognition of task-related features, a greater discrimination between relevant and less relevant features, and a less sequential process of information acquisition. In combination, this suggests that improved diagnostic performance was associated with behaviour that better reflected the utilization of cues in response to task-related problems.

Loveday et al. (2013) concluded that the identification of medical experts on the basis of performance, rather than their experience per se, has the potential to assist in identifying those individuals where additional assistance may be required, either through training interventions, or through the provision of decision support systems. Despite legitimate concerns expressed with over-reliance on pattern-recognition pathways and their inherent weaknesses, including the availability heuristic, anchoring bias, confirmation bias and the framing effect (Scott, 2009), it is likely that providing a safe and reliable learning environment whereby physicians can acquire and refine cues, is likely to deliver safer healthcare in the hospital environment, particularly in those situations where time is of the essence (Aggarwal et al., 2010).

Signature Cues and Cue-Matching Amongst Medical Experts

Expert clinicians 'frame' or contextualise a patient's diagnosis through pattern recognition, utilizing the data acquired through history taking and the initial impression or examination, to develop clinically meaningful representations, with inherent management imperatives. Goffman (1974) illustrated the concept of a 'frame' by using a picture frame as an analogy. A person uses the frame (which represents structure) to draw together a picture (which represents the content) of what is being experienced at the time.

The initial frames are based on matches against cues acquired from previous experience. Inevitably, the initial frame will undergo confirmation or modification as new information is sought or comes to light. Not all cues offered by the patient's history, examination or investigations are of equal utility in modifying the working frame or diagnosis. Indeed, the term 'signature cue' has been invoked to describe a vital or important cue, which is pathognomonic or specifically required for any individual frame.

The signature cue acts as an immediate trigger to seek confirmatory cues for an event or clinical state (i.e. frame). Gonzalez and Brunstein (2009) have described a similar concept of frames or instances composed of three constituent parts (situation, decision, outcome), which inform each other and future similar instances and then lie ready for re-use and potential further modification during a similar event. A study conducted in our intensive care unit utilized cognitive interview techniques to ascertain the cues valued by anaesthetists, surgeons, intensivists and intensive care nurses in determining instability in the post-operative period of cardiac surgical patients who are handed over to intensive care. The outcomes of the study revealed that specific disciplines each held unique cues that were considered specific to their domain, and that there was a positive relationship between years of experience and the frequency of cues nominated by any individual (Festa, de Pont, Crone, Schell, & Wiggins, 2009).

In the case of handover, there is a need to transfer the working frame from one clinician to another. Gadd (1995) described the dual diagnostic theory of consultation, whereby the presenter deliberately attempts to construct a mental model (alternatively, a frame or instance) of the patient's case that is congruent with his or her own. The accuracy of the 'handed over' frame will depend on the presenter's diagnostic and communication skills. For their part, the recipient of the handover uses the presenter's information and communication style to construct a mental model or frame while judging its completeness, lack of ambiguity, and the accuracy of the information provided, as well as the presenter's motivation for including or excluding specific information. Given the diversity of cues that are likely to be utilized by clinicians from different disciplines within a team, it is likely that each member of the team will construct a specific mental model and judgement based on the information conveyed. Therefore, effective communication between team members during a handover allows for the opportunity to detect any ambiguities or missing frame-specific signature cues for any discipline in the newly constructed and now shared mental model, frame or instance.

Evidence of the differences between physicians' signature cues can be drawn from McCormack, Wiggins, Loveday and Festa (2014), who examined the initial visual information-gathering characteristics of qualified specialists and trainee physicians during two simulated clinical deterioration scenarios. In addition to demonstrating different areas of visual interest between the two groups, the results revealed longer fixations by more experienced clinicians and different information acquisition behaviours, depending on the complexity of the scenario. The researchers proposed that the difference in fixation durations, which were even more obvious in the more complex of the two scenarios, was consistent with the proposition that experts and competent non-experts differ in their repertoire of cues or frames in memory. Differences in the areas of interest between the two groups were likely to have been indicative of differing areas of priority when it came to searching for dynamic or confirmatory cues for a given frame or instance.

Conclusion

Accurate and efficient diagnosis remains one of the key skills that less experienced physicians are expected to acquire throughout their training. However, the complexity of the environment within which physicians work, combined with a reliance upon signature cues to quickly and accurate identify a particular complaint, means that the process of developing diagnostic skills must begin at the earliest stages of skill acquisition. The successful utilization of cues and the development of expertise is dependent upon a repertoire of cues in memory, together with a sophisticated mental model that enables patterns of cues to emerge.

An improved understanding of the cognitive basis of diagnosis enables the development of technologies such as decision support and review tools to assist less experienced clinicians with the application of cues and the acquisition and organization of critical cue-based information. Ultimately, as we increase our understanding of expert behaviours and the cues available in an ever-changing healthcare environment, there is an opportunity to inform hospital design and workplace practices, and to optimise training, thereby enhancing medical decision making and patient safety.

Chapter 9

Diagnostic Cues in Major Crime Investigation

Ben Morrison and Natalie Morrison

Decision-Making and Cues

A decision is characterized by three components:

1. a judgement, or choice, between two or more options, resulting in a course of action or commitment, to the exclusion of others;
2. uncertainty, as to the utility of one choice over another; and
3. risk, that one or more options might yield an adverse outcome (Bell, Raiffa, & Tversky, 1988).

Decision-making, or the process by which a decision is formulated, can be described as a subset of an individual's information processing capacity (Howell & Fleishman, 1982; Lehto & Nah, 2006; Wickens & Flach, 1988; Wickens & Hollands, 2000).

Information processing generally involves the acquisition of information, the encoding of information, the recall of information from memory, and the integration of this information to establish a mental representation or an internal explanation or model for how something operates within the external world (Wickens & Flach, 1988). According to Wickens and Hollands (2000), information processing is involved at three key stages of the decision-making process.

The first stage of decision-making involves the perception of external stimuli in the environment. In the second stage of processing, the environmental features attended to are evaluated and integrated, allowing the decision-maker to form a diagnosis of a situation and, if necessary, generate a number of potential responses (Garcia-Retamero, Hoffrage, & Dieckmann, 2007). Finally, in the choice stage, the decision-maker considers the potential outcomes associated with various responses (e.g., their relative degree of risk) and, ultimately, selects a response.

The initial assessment of a situation is a critical aspect of decision-making, since the accuracy of the process determines, in large part, the accuracy of the response. Referred to as situation assessment, it constitutes a significant component of Klein's (1993) model of expert decision-making, and involves a process of matching features in the environment to feature–event associations in memory. It is this repertoire of feature–event associations or *cues* that enables the recognition of a situation as familiar and the application of an appropriate response.

The development of cues in memory is generally dependent upon the accumulation of experience within the operational context (Bishop, 1995). Therefore, the effectiveness with which cues are selected and employed is likely to be moderated by the decision-maker's level of expertise within a given domain. Consistent with this proposition, differences in expert–novice cue processing are evident in a number of domains, including firefighting (Klein, Calderwood, & Clinton-Cirocco, 2010), medical diagnoses (Hammond, Frederick, Robillard, & Victor, 1989), courtroom judgements (Ebbesen & Konecni, 1975), aviation (Stokes, Kemper, & Marsh, 1992), airport customs (Pachur & Marinello, 2013), chess (de Groot, 1966), power control (Loveday, Wiggins, Harris, Smith, & O'Hare, 2013), finance (Hershey, Walsh, Read, & Chulef, 1990), driving (Fisher & Pollatsek, 2007), and nursing (Shanteau, 1991). In these domains, experts are invariably more likely than novices to appreciate the key or signature information associated with a problem scenario (Abernethy, Neal, & Koning, 1994; Charness, 1979; Chase & Simon, 1973a, 1973b; de Groot, 1965; Didierjean & Fernand, 2008). Amongst experts, this results in a tendency to target a relatively limited set of cues to formulate a decision (Jackson, Warren, & Abernethy, 2006; Shanteau, 1992).

In addition to differences in the absolute number of cues available in memory, expert-oriented cues target environmental features that embody relatively greater levels of diagnosticity (i.e., predictive validity) (Schriver, Morrow, Wickens, & Talleur, 2008). The predictive validity of these features is established through repeated interactions with these systems and their utilization in the form of cues tends to be associated with greater levels of decision accuracy (Boreham, 1995; Kirschenbaum, 1992; Morrow et al., 2009; Stokes, Kemper, & March, 1992; Schriver et al., 2008). Moreover, the absence of these cues is associated with degradations in performance, particularly amongst experts, presumably since they develop a significant reliance on these features for problem resolution (Beilock, Wierenga, & Carr, 2002).

As the activation of appropriate cues appears central to proficient decision-making across a number of domains, there are significant implications for lower levels of cue utilization, including the failure to respond to changes in the system state, the failure to respond at the appropriate time, and/or the implementation of an inappropriate response to a change in the system state. Moreover, since expert behaviour is generally associated with greater levels of performance, attempts to improve the performance of less experienced practitioners is likely to be assisted through the examination, identification, and utilization of those cues that are associated with expertise (Wiggins, 2012).

One domain that is particularly vulnerable to shortcomings in training and operational support is major crime investigation. Both in Australia and abroad, there exists a shortage of experienced investigators operating within law enforcement (Deng, 2013). With relatively ad-hoc systems in place, inexperienced investigators may be unable to perform competently while acquiring the necessary skills to advance their level of expertise (Helsen & Starkes, 1999). Given this apparent deficit in investigative expertise, cue-based research offers an attractive approach for supporting the development of diagnostic skills in the domain.

Cues in Major Crime Investigations

In the context of decision-making, major crime investigation is a relatively unique domain since the aim is to identify a potential suspect on the basis of a series of features that are evident as part of a criminal act. For instance, it often requires the capacity to identify key features associated with a crime scene, as well as the capacity to associate these features with the features of the offender (Douglas, Burgess, Burgess, & Resller, 2013).

The notion of linking features of a crime with features of an offender is largely based on behavioural consistency across certain types of offences, and associations between configurations of offence behaviours and offenders' background characteristics (Beutler, Hinton, Crago, & Collier, 1995; Canter, 2000; Canter & Fritzon, 1998; Green, Booth, & Biderman, 2001; Grubin, Kelly, & Ayis, 1997). Salfati and Canter (1999) found evidence for consistency in associations between crimes and offenders across 82 stranger-perpetrated homicides. For example, they demonstrated that murderers who stole non-identifiable property, and acted to minimise forensic evidence left at a scene, were more likely to have had a custodial sentence, and/or to have served time in the military.

The most frequently cited associative model of major crime investigations was developed for the Behavioural Analysis Unit of the Federal Bureau of Investigation (United States). This process, referred to as *Criminal Investigative Analysis*, involves an initial crime scene assessment whereby the investigator seeks out salient crime scene *indicators* (Douglas et al., 2013). For example, Hazelwood and Napier (2004) identified several indicators that are intended to aid investigators in the diagnosis of a staged crime scene (e.g., evidentiary items removed from the victim). Indicators like these, which are generally embedded in features relating to the forensic findings, victimology, and the criminal act itself, will often provide the investigator with information indicative of the offender's characteristics (e.g., age, sex, occupation, residential location, etc.). As indicators trigger the retrieval of associated information from memory, it is presumed that they are largely consistent with the conceptualization of cues in the decision-making literature. Further, it is reasonable to suggest that the number of cues in any given crime scene could be large, presenting a challenge to investigators to identify those features that are of most relevance in deriving inferences about the offender. Therefore, consistent with experts in other domains, expert investigators may be better equipped to recognise and engage the most valid and diagnostic cues available.

Santilla, Korpela, and Häkkänen (2004) tested whether experienced investigators of motor vehicle-related crimes were better able than novices to diagnose links in a series of crimes. Participants were asked to think aloud while attempting to link

10 series of three vehicle-related crimes. The authors reported that experienced investigators who had domain-specific knowledge about car-related crimes performed more accurately compared to inexperienced investigators. In relation to cue-use, experienced investigators consistently focussed their attention on a select subset of features, engaging significantly fewer features during decision-making, compared to their inexperienced counterparts. This is consistent with previous findings that have shown that experts select fewer, more relevant features during information search (Chi, Glaser, & Rees, 1982; Fisher & Fonteyn, 1995; Saariluoma, 1985; Shanteau, 1992). The fact that the experienced investigators were able to divide their attention between a large number of features, and focus information acquisition on a small subset of features, suggests that they engaged schemas based on previous experience and knowledge in selecting information relevant to the task (Bartlett, 1932; Saariluoma, 1990; Schacter, 1989). These findings are consistent with Schraagen and Leijenhorst (2001), who examined the behaviours of forensic scientists searching for particulate material on clothing. The authors noted that experts were better able to mentally reconstruct the actions of the offender at the crime scene, and were consequently superior in focussing their feature search compared to their non-expert counterparts. In their car-crime linkage study, Santilla et al. (2004) also noted that participants who performed more accurately were more likely to use particular features in a problem scenario, and were less likely to use others, in comparison to those participants who were less accurate. This is consistent with experience of cue utilization in aviation, with studies demonstrating that relevant cue-use is associated with differences in performance (Schriver et al., 2008; Wiggins, Brouwers, Davies, & Loveday, 2014).

Significant differences in the use of cues in the context of forensic investigation are also reported by Baber and Butler (2012), who compared novices' (first-year forensic science students) to experts' (experienced crime scene examiners) search behaviours in navigating two simulated crime scenes. The researchers examined participants' behaviour and performance using a combination of a concurrent verbal protocol and head-mounted cameras. Their findings showed that, although both experts and novices considered the likely actions of the offender at the crime scene, novices focussed more on cataloguing features, whereas the experts attended to fewer features that were more heavily weighted in relation to evidential value (i.e., more diagnostic). The authors concluded that the novice investigators approached their search with a view towards detailing the features present at the scene, while experts were more likely to consider evidential associations to the unknown offender as a result of their examination.

Building on Baber and Butler's (2012) conclusions regarding potential differences in the associative meaning of environmental features, Morrison, Wiggins, Bond, and Tyler (2013) used a form of Cognitive Task Analysis to decompose the decision-making processes engaged by investigators during their initial assessment of crime scenes. The researcher's elicited features used by both expert and novice investigators, capturing features relating to the crime scene and, further, features of the unknown offender that may frequently be associated with these crime scene features. Random pairings of scene/offender features were then presented to both expert and novice investigators in a timed paired association task, which was designed to test whether participants recognized an association between the features presented. The results demonstrated a higher level of agreement amongst experts compared to novices in what constituted a valid association. Further, there were significant differences between experts and novices in relation to both the associations recognized, and their speed of recognition. These results provide support for the assertion that expertise is, in part, based on a decision-maker's ability to consistently discriminate between stimuli (Shanteau, 1992; Weiss & Shanteau, 2003). Further, it suggests that experts and novices attend to different information in a scene, and that this triggers different associations in memory (i.e., what they cue). From the evidence reviewed, it is apparent that, consistent with other domains examined in the literature, cue-use differs significantly as a function of expertise in crime-related investigations. Understanding these differences has the potential to steer the development of innovative decision support and training systems in the investigative industry.

Cue-Based Training for Forensic Investigators

By identifying an expert cue set within a specific domain, it may be possible to enable less experienced operators to target a limited number of features that they would not otherwise engage. This is referred to as cue-based training, whereby non-experts are taught to attend to those environmental features employed by experts (Bellenkes, Wickens, & Kramer, 1997; Shapiro & Raymond, 1989; Wiggins & O'Hare, 2003a).

Cue-based training commonly takes two forms, the first of which is based on the notion of *cue discovery* (Klayman, 1988). Cue discovery involves the acquisition of cues ad-hoc through engagement with simulated representations of the operational environment. Through repeated engagement, there is an assumption that users will develop associations between features, sufficient to develop cues, and their acquisition in context will enable the generalization of the cue to the broader environment. The difficulties associated with cue discovery are the costs associated with the development of an appropriate learning environment and the time required for cue acquisition (see also Chapter 7).

The alternative approach involves the presentation of cues systematically, thereby ensuring both their accuracy, and the rate at which they are acquired. However, the disadvantage associated with this approach is that there is a potential lack of generalisability, since individual cues may be acquired in isolation with a lack of understanding of the broader relationships between cues.

Like most training initiatives, establishing the validity of a cue-based training initiative requires an assessment of the application of the outcomes within the operational context (e.g., Wiggins & O'Hare, 2003a). However, in the case of high-stake environments, the contexts within which the outcomes of training courses are evaluated are normally simulated renditions. This reduces the risk of system failures in the event that the appropriate knowledge and/or skills were not acquired.

Although simulated renditions of the operational environment can embody very high levels of visual and functional fidelity, there are difficulties associated with their application for the purposes of cue-based training evaluation. Specifically, they may not present the nuanced features that are necessary to initiate the application of a cue. For example, in an investigation of the cues engaged by experienced practitioners during clinical psychological assessment, Morrison, Morrison, Morton, and Harris (2013) concluded that many of the cues considered critical by practitioners (e.g., client tone, gaze, etc.) were largely absent from the most advanced simulation methods (e.g., virtual patients) available to training practitioners. Therefore, the failure of a learner to initiate an expected response may be due to either the absence of the cue in memory (i.e., the failure of the training initiative) or the failure to activate the cue due to the absence of an appropriate trigger within the simulated environment.

Given the difficulties in assessing performance in some operational environments, together with the limitations of simulation, the outcomes of cue-based training initiatives can be especially difficult to establish, since poor performance might variously be related to the failure to acquire and, therefore, implement cues, and/or the failure to provide an environment that presents the features in a form appropriate to initiate a response. In the case of cue-based training initiatives, an intermediate process of evaluation is necessary that will clearly establish whether cues have been acquired as part of a training process.

Although the acquisition of cues can be difficult to establish, paired association tasks have been used successfully to differentiate the activation of cues amongst experts and novices (Morrison, Wiggins et al., 2013). These tasks have involved the generation and presentation of pairs of features that relate to the concepts under consideration. Respondents are asked to rate, as quickly as possible, the extent to which each of the features is related. Previous research has revealed significant differences in response latency across experts and novices for feature pairs that are considered most related amongst expert investigators (Morrison, Wiggins et al., 2013). Therefore, improving the performance of non-experts involves the capacity to both recognise the relationships between features and to do so within a relatively limited period of time. From this perspective, these cue-recognition tasks offer an opportunity to evaluate the extent to which a cue has been retained in memory, prior to an evaluation of the application of the cue in context.

Although promising, cue-based training does not necessarily offer a solution in industries like law enforcement, for which a significant turnover of workers can be expected, particularly given an increasingly ageing workforce. In these domains, organizations do not possess the time necessary for their workers to fully benefit from extensive cue-based training programmes. In such cases, critical cues may be embedded in Decision Support Systems (DSS), which may serve to develop inexperienced decision-makers while maintaining the integrity of the investigative process.

Cue-Based Decision Support Systems for Forensic Investigators

In the initial stages of a major crime investigation, the volume of information available to investigators is potentially burdensome, giving rise to an increased cognitive load on investigators. The deleterious impact of cognitive load on performance has been well established (Paas, Renkl, & Sweller, 2004; Paas, Tuovinen, Tabbers & Van Gerven, 2003; Sweller, 1988). Consequently, Decision Support Systems (DSS) that are designed to manage cognitive resources during investigations are an attractive prospect. DSS have the capacity to adjust the complexity associated with the performance of a task to a level that matches the skills and capabilities of the decision-maker (Morrison et al., 2010). These adjustments in complexity involve the DSS undertaking one or more of the following stages of information processing: the acquisition of information, the interpretation of information, the integration of information, the comparison between the information and the options available, the comparison between options, the selection of an appropriate option, and/or the implementation of an option (see also Chapter 11).

The utility of reduced processing decision support systems as a learning tool has been demonstrated in both aviation (Wiggins & Bollwerk, 2006) and firefighting (Perry, Wiggins, Childs, & Fogarty, 2012) and might offer opportunities within the context of serious crime investigation. DSS used in criminal investigations may reduce the complexity of information acquired, and assist in the appropriate derivation of offender inferences from the information available to investigators.

Apart from the management of cognitive resources during acquisition and deliberation, DSS have the potential to offer a further level of support by reporting salient associations between offence and offender features. Where clear relationships can be established between features of a crime, inferences can be made from one to another. Here, DSS may generate probability-based information, presenting information relating to the relative frequency of an association between an offence and offender-related features. For example, in New South Wales (Australia), the offender is known to the victim approximately 90 per cent of the time for those homicides involving stab wounds (Chan & Payne, 2013). Therefore, base-rate information presented by a DSS can guide the inexperienced investigator towards the most salient features of an offence that are most relevant in deriving inferences about the offender. Indeed, existing DSS programs such as *Dragnet* have been useful in predicting the residential or base location of serial offenders, based on the location of their offences (see Canter, Coffey, Huntley, & Missen, 2000).

Although it has been argued that probability-based methods of crime investigation may invite error due to their generalization of complex human behaviours (Turvey, 2011), in many cases, the use of an exhaustive process of deduction is often negated by a lack of evidence and the need for a rapid response. For example, for child abduction victims in the United States, the likelihood of retrieving a victim from an unknown offender three days after the abduction is less than 6 per cent (94 per cent of recovered children are found within 72 hours, including 47 per cent found within three hours) (National Centre for Missing & Exploited Children, 2011). Therefore, given the complexity and limitations involved in major crime investigations, DSS which foster a symbiotic combination of probability and investigative expertise represent an additional investigative technique that may prove particularly useful for less experienced investigators.

Conclusion

Although the use of cues is often cited as a prominent factor in decision-making and diagnosis (Jones & Endsley, 2000; Klein, 1993), specific details concerning the nature of cues are often overlooked by researchers. As a result, cues are often qualified as mere information inputs in a larger computational process. However, a vast catalogue of evidence now suggests that cue-use represents a fundamental cognitive mechanism, which may impact decision quality and potentially moderate the differences in performance observed across expertise.

An ageing workforce, and the resultant skills shortage in a range of industries, is making the modelling of expert behaviours a priority in decision-making and human factors research. In domains like criminal investigation, the use of cues offers a significant degree of insight into the development of expertise. Therefore, research that attempts to elicit and model advanced investigative techniques offers a way forward in maintaining and developing investigators' knowledge and skills.

Chapter 10

Diagnostic Cues in Finance

Ben Searle and Jim Rooney

So far, this book has explained how diagnosis can be conceptualized as the extent to which feature–event associations are extracted from the environment to serve as cues for identifying problems and predicting outcomes. Examples have been presented from a range of occupational contexts to show how cue acquisition facilitates context-specific diagnosis. A key conclusion appears to be that – compared to novices or proficient workers – those who are 'experts' (i.e. those who have a high level of ability at forming feature–event associations, and who have also had sufficient time to do so in a particular work context) are superior at accurately diagnosing a situation within that work context. Consequently, experts are likely to make better decisions than non-experts.

This chapter explores some of the boundary conditions associated with this conclusion. While the superiority of experts in diagnosing situations may be generally true, it may be more relevant in some contexts and less in others. The case of the finance industry is examined as a context in which highly-valued expertise in diagnosing risk can have limited impact on the accuracy of financial decision-making.

Off-Task and On-Task Cues

First, let us re-consider what cues are. Earlier chapters focused on cue formation as a learning mechanism that assists in forming diagnostic judgements in relation to specific work problems. If a learned association between a feature and an event assists in the completion of a task, then we could call this a task-relevant cue, or an 'on-task' cue. If a work environment consisted only of features and events relevant to the diagnosis of work problems, the only cues to emerge would be on-task cues.

Yet, no such environment can exist. In reality, work environments contain countless stimuli, including features and events that lack direct relevance to any given task. A sudden burst of white noise from the ceiling may precede an evacuation alert due to a fire drill. A flickering light may precede a storm. Just because people form associations between these features and events does not mean they will necessarily become more capable in the performance of work tasks. We will, therefore, categorise these as 'off-task' cues. Human beings, as social creatures, are especially prone to forming feature–event associations from social stimuli that have little or nothing to do with accurate diagnosis of task-relevant situations, such as off-task communication, interpersonal relationships, displays of incivility, and indicators of social status (e.g. Berger, Cohen, & Zelditch, 1972; Cutting & Dunn, 2002; Mendoza-Denton, Downey, Purdie, Davis, & Pietrzak, 2002). In some environments, these off-task cues may even take priority over on-task cues, such that feature–event associations unrelated to a task will dominate task decision-making.

Social psychologists have long demonstrated that social pressures can influence stated judgements. The conformity studies pioneered by Asch (1951, 1956) are a useful example. Asch showed that, for more than a third of participants, judgements about line lengths were influenced significantly by the erroneous judgements of other group members who were, in reality, confederates of the researcher. Despite the simple nature of the judgement task, the social context of peers making incorrect judgements drove people to conform to the incorrect group norm.

In Asch's (1951, 1956) studies, many participants failed to display accurate on-task diagnosis (correctly judging the length of the line) despite the presence of sufficient task-relevant cues. Instead, a pre-existing off-task cue (conforming with others increases social acceptance and reduces personal discomfort) took priority over on-task cues. There are several explanations for such a phenomenon. If off-task cues are particularly salient, this could interfere with the encoding of on-task features, or such features may be encoded but not processed as deeply as off-task features. Alternatively, on-task features may be processed sufficiently to activate on-task cues and provide an accurate diagnosis, but these cues are discounted or ignored. An accurate task diagnosis would, thus, be ignored as a consequence of having diagnosed the 'appropriate' response for the social context. This would be more consistent with Asch's findings, given that the on-task information was too salient and simple to be easily ignored or misunderstood. Moreover, Asch's studies also showed that when just one of the confederate group members chose the correct response, this inconsistency in off-task feature–event associations was sufficient to encourage nearly all participants to choose the correct response as well. While these all represent possibilities that can be investigated empirically in relation to cue formation, the last is perhaps the most critical, since it suggests that even where an expert operator accurately diagnoses a situation, off-task cues can still determine the task outcome.

Other social influence studies have shown that the greater the ambiguity surrounding a judgement, the greater the conformity to a social norm. For example, Sherif (1935) showed that when participants were exposed to a fixed light point in a darkened room, the autokinetic effect of perceiving the light to be moving was strongly influenced by social pressures – participants' judgements about the extent of movement consistently conformed with norms established by others in the environment. Consequently, we might assume that, in a work context where experts can diagnose problems and predict outcomes with a high degree of accuracy, such experts might be less influenced by social pressures than when outcomes are harder to predict and the accuracy of diagnoses cannot be quickly or easily assessed. Conversely, where diagnosis comes with a high risk of error, even experts may be strongly influenced by social factors.

Under what other circumstances might off-task social cues take precedence over on-task cues? From the example above, we can consider social norms to be a contextual factor that might influence an operator's adherence to on-task cues. Since social norms emerge as a function of culture and values, we will examine these factors together.

Culture, Values and Norms

An organizational culture (or, indeed, any culture) can be thought of as the specific set of *values* and *norms* shared by people within a given context (Schein, 1996). Values are enduring beliefs that some behaviours, goals or achievements are preferable to the alternatives (Rokeach, 1973). For example, in some organizational cultures, product/service quality is valued above other considerations, while other cultures may prioritise customer service or employee quality of life.

Norms, on the other hand, refer to the patterns of behaviour undertaken in a particular context, reflecting both explicit and implicit rules and attitudes regarding what people ought to do and say (Deutsch & Gerard, 1955). An individual's perception of the prevalence of a behaviour among others in a given situation is known as a *descriptive norm* (Cialdini, Kallgren, & Reno, 1991). An individual's perception of how much a behaviour is approved within a given culture or group is known as an *injunctive norm*. Information on the values and norms of an organization's culture may be explicit, but they are often implicit (Payne, 1991; Schein, 1996) and must, therefore, be learned through the detection of cues. In a situation where many behaviours could be performed, and/or where the descriptive norm differs from the injunctive norm, it is less clear how an individual will behave because some competition between norms may occur (Allison, 1992). The focus theory of normative conduct (Cialdini et al., 1991) proposes that the norm most likely to influence behaviour in such situations is the norm that has been made most salient in the situation.

If the central values of an organizational culture involved accurate task diagnosis, and if these values were supported by the organization in terms of extrinsic motivators (status, promotions, financial rewards, etc.) as well as being internalized and displayed by all or most staff, there would be little competition between norms. Rather, there would be strong descriptive and injunctive norms for investing effort into diagnostic accuracy. Employees would feel compelled to devote their attention to extracting task-relevant cues from the environment and give top priority to the implications of such cues.

A hospital aspires to the type of cultural values described above, particularly within units such as intensive care or accident/emergency. Staff are expected to recognise that the highest priority is the accurate diagnosis (and effective treatment) of those patients with the most life-threatening illnesses. Behavioural norms within a hospital often dictate that attending to the diagnosis and treatment of patients is prioritized over personal needs, such as working conventional hours or taking regular work breaks (e.g. Gjerberg, 2003), or organizational needs, such as financial performance. Social status and other extrinsic motivators within hospitals are associated (in principle, if not always in practice) with the capacity to effectively diagnose and treat those patients in greatest need. It is, therefore, not surprising that a cue-based approach to learning, expertise and task performance is particularly useful within a hospital context. Similar arguments could be made of other environments where poor diagnosis can lead to rapid and catastrophic failures, such as aviation.

However, not all organizations operate by these values. Indeed, many expect employees to give highest priority to an organization's financial outcomes, such that the norm of accurate diagnosis within work tasks may be in competition with other powerful norms. Explicit and implicit motivating factors within the organization may reward those who act in their organization's short-term financial interests, rather than maximizing diagnostic accuracy, reflecting a weak injunctive norm. It may be that relatively few employees focus their efforts on diagnostic accuracy alone, and some may overtly prioritise other outcomes (such as organizational profitability), reflecting a weak descriptive norm. Thus, in such organizational cultures, experts may not, in practice, always display more accurate situational diagnosis than novices or proficient employees.

Moreover, conceptualizing organizational systems requires an understanding of the multiple levels involved (Hitt, Beamish, Jackson, & Mathieu, 2007). Work tasks are often nested within job and/or team responsibilities, which are, in turn, nested within organizational goals. Taking this perspective, we should consider how even when a task-level evaluation of cues suggests one diagnosis and consequent action, there is the potential for conflict with team responsibilities or organizational goals. For example, in the process of a sales conversation, a salesperson may diagnose (on the basis of learned feature–event associations) that the customer is likely to need only the cheapest of the products available. However, this diagnosis may not dictate which product the salesperson attempts to sell to the customer, since personal, team and organizational goals (such as sales targets) are likely to also to influence this decision. Indeed, the process of cognitive dissonance (the discomfort felt by someone holding two or more inconsistent cognitions; Festinger, 1962) could lead the salesperson to revise the diagnosis to be more consistent with sales targets (i.e. the customer must need a better product).

The example above not only highlights how higher-level goals can affect cue-based diagnosis leading to less accurate task-based decisions, it highlights how motivational factors can affect this process. Research has long shown that decision-making can be influenced by organizational goals as well as incentives, expectations and control systems (O'Reilly, 1983). In addition, social identity theory suggests that when one's social identity (e.g. as a team or organization member) is more salient, one is more motivated to act in a manner that supports the needs of that group (van Knippenberg, 2000). So, once again, social cues unrelated to diagnosing a specific situation have the potential to influence diagnosis and consequent action.

Cues in Credit Underwriting

To illustrate these points, this chapter describes the work of credit underwriting (loan decision-making) for residential housing purposes (hereafter, home loan credit decision) in which cultural, social and organizational factors may influence decision-making above and beyond the application of expertise to task diagnosis. To provide a reasonable measure of understanding of the environment facing experts responsible for home loan credit decisions, we need to start with a brief outline of the work context from both an economic and social perspective.

While the key diagnosis and decisions addressed here are, in broad terms, applicable to similar home loan markets globally, those features that are specific to the Australian context will be highlighted throughout this chapter. It is not intended to provide a comprehensive global comparison but to help foster a greater appreciation of on-task/off-task factors. Following an overview of the environment at an institutional level, we progress to current insights on the individual consumer choice of mortgage loan before outlining the events and features relevant to the credit decision-making process involved in the approval of home loans.

The Australian Home Loan Industry

The home loan industry is a significant part of the financial services sector in Australia (which, in total, is the third largest sector of the Australian economy), contributing 7.5 per cent of GDP and 3.6 per cent of total Australian employment respectively (Euromonitor, 2009). Starting with the economic context, the industry is dominated by five large domestic banks, accounting for 76 per cent of its total home loan assets and approximately 90 per cent of new home loans (Euromonitor, 2009). Until the advent of the Global Financial Crisis (GFC), non-bank financial institutes had been an emerging threat to traditional banks in some markets as they offer substitute products at competitive prices. This aspect is addressed later in the chapter as it influences both feature/event linkages and on-task/off-task factors.

A detailed discussion of home loan products is beyond the scope of this chapter. However, in general terms, a home loan (the product of interest in this chapter) consists of the following attributes:

1. Targets individual citizens or permanent residents who wish to use the loan to purchase, build and/or renovate residential property. The key purpose is usually based on a plan to live in (owner occupied) or rent to others for the purpose of receiving income (investment or buy-to-let). Secondary purposes have been increasingly important in recent years including

      funding for house renovation, for the establishment of private enterprises, or for investment in financial instruments (Kamleitner & Kirchler, 2007).

2. Loan funds are provided only to such individuals who meet the financial institution's lending decision policy/criteria and/or are judged by qualified credit underwriters (also known as credit officers or underwriters) to be eligible for the loan. The features associated with this aspect are discussed later.

3. The products can embody a number of different characteristics associated mainly with different forms of funds access (including for options for additional funds), repayment (including options to defer regular repayments) and discharge/closure.

4. The accepted collateral or security for home loans is usually standalone or semi-detached residential houses or multiple-occupancy residential buildings (such as flats or condominiums) that meet specified lender or mortgage insurance standards and are independently valued as being within the required standards of the financial institution.

5. The pricing of home loans (in common with most other loan products) is predominantly based on the charging of interest on the loan amount borrowed or outstanding – at a fixed (for an agreed period) or variable (can change after due notice) rate. However, in recent decades, there is an increasing trend to lower rates that are offset (from an institutional perspective) by fees triggered by processing activities such as administration/account keeping, or the availability/use of optional features. For example, Australian home loan providers usually charge an exit fee for the discharge of fixed rate loans before the end of an agreed fixed rate period (known generally as break fees).

Activities that ensure compliance with these characteristics are considered to be the most pertinent of the on-task factors associated with the home loan product. This can be observed both from the academic literature (e.g. Bourke & Shanmugam, 1990) and from practitioner artefacts, in particular, the credit policy documentation specific to each lending institution. A brief examination of the individual consumer product choice context will provide further insight into other applicable on-task factors and introduce some of the key off-task factors.

Consumer Selection of a Mortgage Product

There has been a good deal of research into the factors that influence consumer choice of loan products and related loan finance decisions. In economic terms, the demand for home loan products is directly linked to the general economic environment, through factors such as the cost of debt, employment levels, growth in asset prices and the availability and terms of credit (Montgomerie, 2006). Research on this aspect ranges in focus from economic modelling (e.g. use of options analysis; Daglish, 2009) through to experiments based on concepts taken from the literature on cognitive psychology (e.g. use of the mental accounting concepts developed by Kahneman & Tversky, 1979; see also a study by Ranyard, Hinkley, Williamson, & McHugh, 2006).

While research on loan selection by consumers tends to focus on economic determinants, embedded relationships have also been examined in both the academic and marketing practitioner literature. Strong, embedded relationships have been demonstrated to shift customer motivations away from the pursuit of immediate economic gains towards the enrichment of relationships through trust and reciprocity (Powell, 1990; Smitka, 1991). Consistent with this finding, the importance of establishing trust (or confidence) in the bank–customer relationship has also been discussed in the bank marketing literature (e.g. Gill, Flaschner, & Shachar, 2006; Hawke & Heffernan, 2006). Finally, there have been several studies on customer loyalty in banking which, until recently, have concentrated on consumer intentions, rather than actual behaviour, and the measures of loyalty have been orientated according to this perspective (Bauman, Burton, Elliott, & Kehr, 2007).

The work context examined in this chapter is focused on consumer access to an opportunity to submit a home loan request or application. This access occurs through one of four key distribution channels: retail bank branches; internet banking; telephone banking; and third party brokers. Of these channels, mortgage brokers – independent sources of mortgage product sales and service – appear to be most influential in terms of off-task influences on credit decisions over recent years. A 2003 survey of broker-originated mortgage lending by the banking industry regulator in Australia identified a total of 56 institutions that use brokers to originate mortgage loans. This represents 25 per cent of Australian institutions by number, but 96 per cent of institutions on an asset-weighted basis.

The survey identified a number of positive aspects of the most common practices regarding broker-originated lending, namely: the use of formal broker accreditation processes; the use of automated systems to track broker-introduced loans; an internal review of broker-originated loans with records of rejected broker-introduced loan applications; and a rejection of lending business from non-contracted brokers. However, the survey also identified several negative aspects of broker-originated lending. Firstly, payment of broker remuneration was solely determined by the volume of business generated. Based on average origination and trailing commissions on housing loans, the Australian Prudential Regulatory Authority estimated the value of total commissions to be 1.14 per cent of loan value, which equates to over $800 million in commissions paid by banks to brokers in 2003 (Chanthivong, Coleman, & Esho, 2003). Despite this, recourse to the broker if a loan is in default appeared to be limited, and credit limits on exposures to individual brokers or aggregate broker-introduced business were also quite limited.

Statistics indicate that credit approval rates on broker-introduced loans are generally lower than non-broker-introduced loans (Chanthivong et al., 2003). There is limited time-series data on whether broker-originated loans are higher risk. However, a large number of institutions do not report higher impaired assets rates on broker-originated loans. This may be due to the inadequate tracking of broker-originated loans, economic conditions and/or the relatively recent introduction of brokers at some banks, although anecdotal and proprietary evidence suggests otherwise.

Decision-Making Processes in Home Lending

The approval and valuation of home loans can be a challenging task for financial institutions whose aim is to devise and implement risk management regimes that balance risk (e.g. losses from loan repayment default and foreclosure) and reward (e.g. loan interest and fee income). It is a complex problem from both a theoretical and an empirical point of view, since a fair valuation associated with a credit decision 'should' take into account a range of risks including, but not limited to, interest rates. This range of features can materially impact the two events on which this process is focused, thereby preventing future credit default, and/or losses associated with an approved loan.

The literature relating to loan credit risk and decisions has tended to focus on loan performance after the decision has been made, rather than the decision-making process itself. This can be explained largely by a preoccupation among finance researchers with economic explanations, relating to the prediction and/or measurement of repayment default performance for approved loans, ex ante the approval process (Altman & Saunders, 1998). Nevertheless, there is some limited research based on the behavioural decision-making literature. However, this also tends to focus on loan performance or is limited to specific aspects such as the impact of lending experience on the consistency and speed of the decision-making process (e.g. Andersson, 2004). The behavioural research involves the study of cognitive decision-making processes with an emphasis on the identification and processing of decision cues (including the speed of identification and the selection of contradictory data as well as the use of mental checklists).

To aid in the interpretation of decision-making research in the home loan approval context, it is important to understand the key procedural steps that are integral to the lending decision for home loans. Based on Bourke and Shanmugam (1990) and modified for home loans (rather than commercial loans), there are eight key stages:

1. Acquire basic information about the borrower – including historical and current bank statements, as well as general background information, usually in a prescribed format that specifies minimum data requirements such as full name and current/recent employer details.
2. Acquire basic information about the loan request – including the size and purpose of the loan and sources of loan repayment.
3. Conduct a preliminary review of the risk – involves the assessment of various on-task features that are influenced by a combination of the prescribed

and documented credit policy,[1] procedural training, and individual credit underwriter experience in evaluating the risk of default and/or loss. The key features that need to be assessed can be summarized into three, not entirely independent, categories:

- the individual applicant's willingness to repay the mortgage;
- the individual's ability to repay, incorporating features outside their personal control such as general economic conditions; and
- the recoverability of loaned funds recoursed through the sale of the mortgaged property(ies) should the individual default on repayment.

The result is the determination of an initial risk rating with confirmation of apparent adherence to documented procedures to guide the rest of the process.

4. Acquire more complete information – includes additional detail of the proposed loan in response to queries generated in the earlier steps, such as details on income/employment; financial commitments and plans; and other supplemental financial information required for a final decision.

5. Verify the critical information – including routine bank and employment checks and copies of various proofs of identity, employment and residential stability (e.g. utilities bills, etc.).

6. Refined the analysis of risk – including on-task features such as establishing the commitment of applicants, assessing applicants' financial success and determination to pay loans; judging external risk/future (e.g. economic, political, etc.); identifying and weighing financial strengths and weaknesses; evaluating expectations for current and future income; evaluating, projecting and estimating the strength of cash flow in relation to all demands (including debt servicing of requested loan); estimating current or potential pressures on financial needs and liabilities; and evaluating secondary protection of assets (including the value of any new real estate collateral).

7. Make the loan final decision and offer – including the assignment of an overall risk rating assessment of the likely occurrence of either or both of the two events and the attachment of any conditions of offer over and above those applicable to the standard specification of the home loan product offered to the particular applicant. If the rating is unacceptable, the process ends with notification of a 'not approved' decision.

8. Structure the loan – including confirmation of the appropriate loan product, maturity and pricing; seeking appropriate acceptance of any conditions

---

1   As documented in a formal document usually titled Credit Policy or Lending Policy specific to that particular financial institution. It is almost always created by a separate Risk Management function that is usually independent of the credit decision-making function and provides loan decision guidelines.

by the customer; and documenting/closing the loan in accordance with approval.

Feature–Event Association: 'Real Life' Credit Decisions

While the list of process steps for home loan credit decision-making reflects industry practice at an abstract level, it is largely idealized, prescriptive and based on assumptions of consumer and employee decision-making rationality. These assumptions are drawn largely from the outcomes of survey-related research in which participants are asked to consider hypothetical or abstract scenarios in acquiring and interpreting information. For example, Gibson (1983) used a questionnaire to identify the financial ratios that are considered important by the 100 largest banks in the United States in making lending decisions to commercial customers. There are benefits associated with this type of survey research, insofar as it provides a level of abstraction that assists in understanding the decision-making process. However, these methodological approaches can yield inconsistent outcomes, particularly in the use of higher-order cognitive processes and reliance on memory (Nisbett & Wilson, 1977; Pearson, Ross, & Dawes, 1992). There are also issues associated with the social desirability of responses and other forms of common method variance that can emerge through survey-based research (Podsakoff, MacKenzie, Lee, & Podsakoff, 2003).

As a consequence of the limitations of previous research, it appears that the rich detail in the variation of individual decision-making preferences and processes may have been overlooked. Indeed, Carruthers (2012) notes, in referring to assumptions associated with lending practices, that, '[In the] ... messy real world, all of these assumptions get violated' (p. 741). This is particularly evident in research that demonstrates that lending decisions can differ between individuals, even when the information provided has been identical (Innes & Lyon, 1994). Examples of the rich detail around which individual differences might be evident include the evaluation of on-task features such as:

- the impact of the new loan repayment commitment on the current personal finances of the applicant;
- experience with the effects of a range of micro-economic variables such as residential and employment stability;[2]
- the incorporation of a broad range of political and macro-economic risks;
- the evaluation of expenses and related outgoings based on track record and/or sustainable cash outflow indices for equivalent household units (e.g. Henderson Poverty Index);
- the anticipation of future cash flows and capacity to service debt; and

---

2   The experience, often embedded in Credit Policy as well underwriter experience, that customers who have longer periods at one residential address and with one employer, tend to be more willing to meet loan repayments on the date due.

- the evaluation of personal assets and liabilities incorporated with the requested loan.

Further exploration of the impact of these features is warranted, focusing on the 'real life' use of task cues, as well as the key factors that impact the process. While there has been some research that has incorporated the contextual nature of the decision-making environment and the contingent nature of the process, it is these processes that potentially explain the on-task and off-task influences on the decisions that are formulated by individual credit underwriters.

Individual Factors Influencing On-Task and Off-Task Cue Utilization

One feature that has been examined in some detail in the context of financial lending is the influence of underwriter expertise on decision-making behaviour. For example, Rodgers (1999) compared a sample of 40 experienced credit officers with 67 inexperienced MBA students on their judgement of the importance of conflicting information provided by a loan applicant. Consistent with other research on this subject, experts searched for different cues, particularly where the information required to make a decision was conflicting. Experts engaged in a different process of information acquisition to arrive at a lending decision from that used by less experienced subjects. However, this did not appear to result in consistent credit decisions among the experts.

In a more recent study focused on consumer lending, Andersson (2004) confirmed the lack of consistency on decision outcomes amongst lending experts. Contrary to prior research, lending experts searched for significantly more features than less experienced subjects in formulating their decision. There are three potential explanations for this outcome, including that the task characteristics associated with the 'naturalistic situations' alerted more experienced subjects to search for additional cues (a research methodology issue), the absence of search cost for this additional search activity, and/or a combination of risk aversion and perceived responsibility.

Regarding the last of these, the conservatism of credit underwriters, Innes and Lyon (1994) suggest that there is a greater sensitivity to negative data than to positive information. This is broadly consistent with Kahneman and Tversky (1979), and this tendency seems to extend to how the source of the information is considered (Beaulieu, 1994). For example, Shanmugam and Bourke (1992) found that credit underwriters were willing to change their decision when reliable third party information was introduced.

Consistent with the notion of reliability, Ruchala, Hill and Dalton (1996) have demonstrated that a credit underwriter is more likely to approve a subsequent loan if he/she has approved a loan for that customer in the past. Similarly, McNamara and Bromiley (1999) considered the assessment of management credit risk, and concluded that 'they are the outcome of a complex set of interactions played out over time between lenders and borrowers' (p. 228). Finally, Cole (1998) reported that a prior relationship is more likely to lead to loan approval, although the duration of that relationship appears to be unimportant. Petersen and Rajan (1995) explain the nature of this relationship as a reduction in information asymmetry which results not only in a high likelihood of credit approval, but also in a reduction in both loan pricing and looser loan contract conditions, including levels of loan collateral and the availability of credit (Harhoff & Körting, 1998).

In the context of decisions relating to commercial loans, 'gut feel' appears, anecdotally, to be an important determinant in the decision process. For example, Jancowicz and Hisrich (1987) interviewed 20 commercial lenders across four US banks, developing 132 decision constructs that were subjected to repertory grid analysis. This analysis identified eight key decision types and, in particular, identified intuition as a key feature. In the absence of decision support tools, intuition is presumably intended to reduce the cognitive load that is normally associated with the decision-making process. Consistent with this proposition, Casey (1980) assessed the decision-making process of bank officers in the United States and noted that increasing the amount of information available reduced both decision time and predictive accuracy.

In comparisons between different applicant loan profiles, Biggs, Bedard, Gaber and Linsmeier (1985) noted that amongst experts, where they were similar, there was an increased reliance on compensatory strategies. However, non-compensatory strategies tended to be adopted when the profiles were dissimilar. This suggests that experts simplify their decision-making processes under higher cognitive load, which may then impact the consistency and accuracy of subsequent judgements. Anecdotal support for this finding can be drawn from the events that preceded the GFC.

Despite initial forays into the nature of credit decision-making, the existing research tends to fracture, and fails to take into account the complexities associated with the process. For example, there has been very little research concerning the role of emotions and mood on credit decisions (Elsbach & Barr, 1999; Fineman & Sturdy, 1999). Similarly, there is a lack of understanding of the specific cognitive models that are employed by credit officers, the extrinsic and intrinsic factors that motivate their behaviour, the effectiveness of training, and the impact of loan pricing on credit decision-making. While not exhaustive, these gaps in our understanding potentially highlight some of the future research priorities that might be directed towards modelling on-task/off-task diagnosis.

Culture and Normative Factors Influencing On-Task and Off-Task Cue Utilization

One of the potential extrinsic influences on the individual loan decision-maker is the institutional culture present at both an industry and corporate/firm level. Anecdotal evidence suggests that Australian financial institutions are generally regarded as overly conservative and particularly focused on the quality of loan collateral (Nijskens & Wagner, 2011). As Sarasvathy, Simon and Lave (1998) observe, 'bankers use target outcomes as reference points and operate by attempting to control risk within the existing structured problem spaces, avoiding situations where they risk higher levels of personal responsibility' (p. 218). While there is a large body of research concerning organizational culture, there is a need to investigate how specific sub-cultures and perceptions of power and influence might impact decision-making processes at the level of the individual.

The increasing use of third party sales brokers, at least anecdotally, appears to exert implicit pressure to approve individual loans. The 2003 APRA Survey (Chanthivong et al., 2003) identified that commission-only remuneration was a significant negative aspect of broker-originated lending practices. A consequence is that the additional pressure from individual credit underwriters, as well as the decision-making process as a whole, appears to increase the proportion of loan applications that are approved. Therefore, it is important to establish the role of explicit and implicit coercion that may emerge from cultural differences between organizations and whether this impacts both the assessment of applications for credit and, ultimately, credit-related decisions.

Even within an organization, there appear to be apparent differences between the requirements for documentation and subsequent norms associated with the use of this information in the context of approvals for loans. For example, McNamara, Moon and Bromiley (2002) observed that documented processes can be over-ridden by perceived institutional priorities, and particularly the drive for increased profitability. In the case of McNamara et al. (2002), the result was a bias towards more favourable treatment of new lending customers. As Sutcliff and McNamara (2001) note, 'risk rating analysis' means that 'decision-maker behaviour is situated and is not simply a function of individual choice, Rather, in organizational settings decision-makers are subject to a hierarchy of influences that affect the decision processes they use and their resulting decision choices' (p. 484).

Conclusion

In recent years, loans for credit have become as much a moral issue as they have a cognitive process. The moral hazard was brought into sharp relief by the debate on the role of credit default options along with related financial derivatives in the lead-up to the GFC. As many authors and media commentators observed, there were a number of alternatives to 'pass off' the consequences of a bad credit decision, not all of which turned out to be robust. However, in more 'normal' economic times, these alternatives are pervasive whether they be a temptation to rely on external professional advice in the form of property valuations; a reliance on mortgage insurers to carry shortfalls in loan loss recoveries; internal and external economic outlook; and/or mortgage-based investment securities. While this is an important consideration, there is a more immediate need to consider the cognitive perspective and the implicit impact of off-task factors on credit lending diagnosis and decision-making (Belsky, Case, & Smith, 2008).

A closer examination of how credit decision-making is influenced by cultural/ normative factors, particularly within credit functions but also within financial institutions in general, would be valuable if policy-makers desire a systemic change to the underlying risks of credit-based economic crises. Indeed, a comprehensive exploration of the way that explicit organizational policies and procedures influence credit approval processes would be an important step in the right direction. The experienced power of credit officers to apply these policies and to resist direct and indirect pressures to approve risky loans may be influenced, to some degree, by their formation of off-task feature–event associations linking more approvals with better outcomes.

Chapter 11
Diagnostic Support Systems

Nathan Perry

A combination of factors such as an ageing workforce and growth in emerging international economies is resulting in substantial increases in the demand for operators in many high-reliability domains. For instance, in the aviation industry, it is estimated that 498,000 new commercial airline pilots will need to be trained over the next 20 years (Boeing Commercial Airplanes, 2013). An issue associated with a workforce comprising operators who have not accumulated sufficient experience to have developed expert cognitive processing skills, so called less-experienced operators, is the increased potential for error due to limitations in human information processing.

Emergency situations present a particular problem for less-experienced operators. The significant demands that are placed on cognitive processing resources to develop a situation diagnosis within time-critical situations provides the potential to overwhelm the processing capacity of inexperienced personnel. Consequently, to ensure that system integrity is maintained, there is a need for the development and application of systems that have the capacity to aid the diagnostic ability of less-experienced operators. Using an aviation accident as a context, this chapter provides an overview of the information processing limitations of less-experienced operators and discusses a potential approach for supporting situational diagnosis during complex and time-constrained situations.

The Diagnostic Failure that Crashed Air France Flight 447

On 31 May 2009 at 8:29 p.m., Air France Flight 447 (AF447) departed Rio de Janeiro Airport bound for Paris (Bureau d'Enquêtes et d'Analyses, 2012). Three and a half hours into the flight, the captain left the flight deck for a rest break, leaving the aircraft in the charge of two first officers. Before departing the flight deck, the captain appointed the much less experienced of the two first officers in command of the Airbus A330–203 who performed the role of the pilot flying (PF). Shortly after the captain left for his break, the aircraft entered the edge of a tropical thunderstorm. Icing conditions were encountered, resulting in the accumulation of ice in the pitot tubes causing the indicated airspeed readings to become unreliable and the disconnection of the autopilot.

The PF took control of the aircraft while the pilot not flying (PNF) became fixated on identifying the cause of the autopilot disconnection. The PF struggled to control the aircraft's bank angle in the turbulent conditions, a task that likely taxed his cognitive resources. For some unknown reason, the PF provided a continuous nose-up input on the control sidestick, causing the aircraft to pitch up and gain altitude. The continued climb resulted in a pitch-up attitude that approached an aerodynamic stall, triggering the stall warning. The pilots appeared confused by the warning and the approach to stall situation was not recognized. The accident investigation suggested that, due to the PNF's focus on the error messages, he was likely unaware of the aircraft's rate of climb, while the cognitive resources of the PF were devoted to controlling the roll of the aircraft.

The captain was called to the flight deck as the PF continued to make pitch-up inputs on the sidestick and the stall warning sounded once again. Despite the stall warning, the pilots failed to appropriately diagnose the situation and the aircraft continued to climb, despite the appropriate recovery procedure requiring a pitch-down attitude. The angle of attack of the aircraft continued to increase as the airspeed decreased, eventually resulting in an aerodynamic stall that was not diagnosed by the pilots. The aircraft was rapidly descending in a stall when the captain returned to the flight deck, but he experienced difficulty piecing together the information and also experienced confusion. The aircraft eventually crashed into the ocean, killing all 228 crew and passengers on board.

From the description of the accident, it is clear that a number of complex factors contributed to the occurrence, including unreliable airspeed indications, aircraft design issues, and ineffective crew resource management (Bureau d'Enquêtes et d'Analyses, 2012). However, it seems that a misdiagnosis of the situation, particularly by the two first officers in control of the aircraft, may have played a major role. During much of the event, the PNF entrusted the comparatively inexperienced PF with the handling of the aircraft, while the PNF attended to the error messages. However, the evidence suggests that the cognitive resources of the PF were severely taxed by the task of controlling the aircraft in the turbulent conditions, which left little cognitive capacity for the evaluation and integration of the available information to form an accurate situation diagnosis, particularly for relatively inexperienced operators (Bureau d'Enquêtes et d'Analyses, 2012).

The crash of AF447 highlights the vulnerability of human cognitive processes to error, particularly in highly complex and time-constrained situations. Under these conditions, the cognitive resources necessary for the acquisition and integration of information to develop a good mental model are likely to be severely stretched. Consequently, it cannot be assumed that human operators will necessarily possess the cognitive capacity to process information to form an accurate situation diagnosis.

Information Processing and Less-Experienced Operators

One of the hallmarks of expertise is an ability to develop an accurate diagnosis of a situation through recognition (Klein, 1989). Through the accumulation of experience in a vast array of different situations, experts learn the associations between features present in the situation and the likelihood for different events (Wiggins, 2006, 2012). Repeated exposure to particular feature–event associations results in the formation of cues that can be used for the efficient retrieval of information from long-term memory. Once retrieval cues are established from repeated activations of the feature–event pathways, the presence of key or signature cues in a situation can lead to the automatic retrieval of the associated information from long-term memory (Ericsson & Kintsch, 1995).

Evidence of experts' superior recognition abilities can be drawn from Chase and Simon's (1973b) study with chess masters. After viewing the positions of chess pieces on a board for a period of five seconds, both novice and chess masters were asked to reproduce the board configurations. For board configurations that were likely to be experienced in a game of chess, chess masters could reproduce the positions on the board more accurately than the less skilled players. However, when random board configurations were tested, the performance of experts was no better than the performance of the novices. This evidence suggests that the experts were relying on the recognition of previously encountered board positions that was acquired from vast experience.

The information processing advantage for experts is that they can engage top-down processing. The capacity to draw upon previous knowledge and experiences from long-term memory activates relevant schemas that enable information acquisition to be directed towards features that are known to be vital to the situation (Shanteau, 1992). In contrast, operators who have not accumulated sufficient experience to have acquired memory retrieval cues are more likely to engage intensive bottom-up processing when assessing a situation (Sohn & Doane, 2004). With a reduced capacity to rely on the automatic activation of information from long-term memory, less-experienced personnel tend to have greater difficulty distinguishing relevant from irrelevant information, resulting in a greater proportion of information being processed in working memory (Lipshitz, Omodei, McLennan, & Wearing, 2007; Wiggins, Stevens, Howard, Henley, & O'Hare, 2002; Sohn & Doane, 2004).

The influence of 'experience' on cognitive processing is supported by research into the way that experienced and less-experienced pilots arrive at a situational diagnosis (Sohn & Doane, 2004). In particular, the ability for experienced pilots to accurately diagnose the current state of an aircraft, and project its state in the near future, can be predicted from their skill in retrieving information from long-term memory. In contrast, situational diagnosis amongst less-experienced pilots tends to be better predicted by their working memory capacity. This suggests that less-experienced pilots rely on their ability to process information in working memory when forming a situational diagnosis (Sohn & Doane, 2004).

In routine circumstances, the elevated levels of cognitive load that results from the evaluation of information in working memory will not necessarily have a detrimental impact on performance. However, the issue associated with a high working memory load is that performance is subject to working memory capacity limitations. Theoretically, situational diagnosis will be impaired when the information load exceeds the operator's working memory resources (Endsley, 1995a; Tsang & Vidulich, 2006). Not only does a high working memory load limit the resources available for processing additional information, but the ability to extract relevant from irrelevant information is also degraded (de Fockert, Rees, Frith, & Lavie, 2001). Consequently, the significant demands that complex and time-constrained situations place on cognitive processing resources, such as the emergency event experienced by the pilots of AF447, can seriously threaten both the accuracy and the efficiency of situational diagnosis amongst less-experienced operators (Bureau d'Enquêtes et d'Analyses, 2012). By virtue of their tendency to process a greater proportion of information in working memory, the ability for less-experienced operators to identify and integrate the information necessary for situation diagnosis can be severely impaired (Hahn, Lawson, & Lee, 1992; Svensson & Wilson, 2002).

With the high cognitive demands that can be faced during emergency events, it cannot be assumed that human operators will always be able to easily identify and integrate the cues necessary to appropriately diagnose a situation. Less experienced operators, in particular, are at an increased risk of performance degradations due to increased cognitive overload. Therefore, there needs to be some consideration of the design of systems to assist operators to appropriately process information and thereby develop an accurate diagnosis of the situation.

Supporting Diagnosis in Less-Experienced Operators

High-reliability domains often require operators to rapidly develop an accurate diagnosis of the situation. However, the cognitive load that less-experienced operators face in such circumstances can constrain this process. The accident involving AF447 demonstrated that operators' cognitive resources can become constrained to a point where they are unable to mentally integrate the information necessary to form an accurate situational diagnosis. Therefore, the potential exists to support situation diagnosis amongst less-experienced operators by reducing the level of information processing that must occur in working memory, thereby improving diagnostic accuracy.

A review of the cognitive processes involved in situational diagnosis provides insight into potential sources of increased cognitive load that could impair the diagnostic performance of less-experienced operators. If the situation cannot be recognized through the activation of signature cues, a more top-down process of situational diagnosis is likely to be engaged. First, the operator must acquire salient information, which is then integrated and interpreted in working memory for a mental model to be developed of the situation (Stokes, Kemper, & Kite, 1997). Further information acquisition can then be directed towards information that can confirm the mental model, with the situation considered diagnosed once the appropriate mental model has been activated. The reduced capacity for less-experienced operators to distinguish relevant from irrelevant information presents the potential to load working memory with non-diagnostic information. With working memory resources expended processing information that is potentially of lower diagnostic value, fewer resources are available subsequently for the integration of relevant features to form an accurate mental model of the situation. Therefore, systems that can facilitate information acquisition and information integration offer the potential to improve situational diagnosis amongst less-experienced operators.

Facilitating the integration of information for the purpose of reducing working memory load is a well-established design goal. According to the proximity compatibility principle, if a task requires the mental integration of multiple sources of information, cognitive processing can be facilitated by locating the information in close physical proximity (Wickens & Carswell, 1995). The close information proximity enables the operator to perceptually integrate and combine features, rather than having to rely on manipulating information in working memory. In fact, evidence has shown that displays that facilitate the integration of perceptual information can improve the speed and accuracy of decision-making on tasks that require comparative assessments (Marino & Mahan, 2005).

Although systems that support the integration of information have been effective in improving performance, further reductions in cognitive load can be achieved through a reduced processing approach that also limits the amount of information available to be processed (Morrison, Wiggins, & Porter, 2010). The reduced processing approach to support system design is based on evidence that highly experienced personnel can make accurate decisions on the basis of a limited number of highly diagnostic cues (Einhorn, 1974; Klein, 1989, 1997; Slovic, 1969). Theoretically, the acquisition of information, in addition to these key cues, does not necessarily improve decision accuracy. In situations where there are significant amounts of information available, it is only by prioritizing critical information that demands on information processing are reduced with a minimal loss of accuracy.

Morrison et al. (2010) investigated the influence of different reduced processing interfaces to support the performance of trainee forensic investigators in formulating decisions regarding the evidentiary value of fingerprints. The interfaces varied in the potential to load working memory through the manipulation of the quantity of available information and the necessity to integrate information in working memory. The reduced processing interfaces provided various levels of restriction over the amount of information that could be accessed on the basis of participants' information prioritization. The level of information restriction ranged from unrestricted information access through to limiting access to the three features that were assigned the highest priority. A processing advantage was evident for decisions made using the reduced processing interfaces. Limiting the availability of information that was considered relatively lower in diagnostic value enabled trainees to arrive at an accurate decision more rapidly. Despite the most restrictive interface limiting access to only the three most critical pieces of information, no loss of decision accuracy was observed.

Perry, Wiggins, Childs and Fogarty (2012) investigated the capacity for reduced processing interfaces in facilitating cognitive processing amongst less-experienced fire officers. The task for the fire officers was to extract information from conceptual decision support systems to make point-of-entry decisions in the context of conducting a search and rescue. Consistent with Morrison et al. (2010), three interfaces were investigated that varied the cognitive processing load. The decision-making performance with a full processing interface was compared against performance with two reduced processing interfaces that enabled information filtering and information integration. Further, the most restrictive interface limited access to only the three pieces of information that the user considered most critical.

Aligned with theories of expertise, experienced fire officers possessed a superior ability to make decisions in time-constrained scenarios by limiting information acquisition to the most critical cues. However, the performance of the non-expert fire officers was improved using the reduced processing interfaces. The introduction of the reduced processing interfaces resulted in faster and more efficient decision-making. Most significantly, reduced processing interfaces reduced the tendency of the less-experienced fire officers to re-access information, thereby reducing the burden on working memory resources. The outcomes of this study suggested that, provided operators have sufficient understanding of the diagnostic validity of the available information, the application of interfaces that reduce the propensity for superfluous cognitive processing can enable a greater proportion of cognitive resources to be allocated to in-depth evaluation of the most critical information (Perry, Wiggins, Childs, & Fogarty, 2013).

The reduced processing systems investigated by Morrison et al. (2010) and Perry et al. (2012, 2013) can be considered a test-bed for an approach to supporting the performance of less-experienced operators by encouraging the processing of the most critical information. Aiding less-experienced operators in the identification and processing of the most critical information, while reducing processing of non-critical information, can result in improved performance. Such a design principle holds considerable promise for complex environments, particularly during emergencies that require information to be processed in a manner that enables rapid diagnosis. The next section considers key issues in the application of reduced processing systems.

Application of Reduced Processing Support Systems

Reduced processing systems offer the potential to assist less-experienced operators to process information efficiently to manage their performance in complex and time-constrained situations. However, discussions concerning the potential applications of these systems have been limited, with research to date focussing primarily on conceptual questions. This section highlights some of the issues that might need to be considered for the application of reduced processing systems within the operational environment.

Central to the concept of reduced processing support systems is the capacity to prioritise information as an event emerges. The intention of reduced processing support systems is to limit the presentation of information to only the most diagnostic information, thereby reducing the potential for superfluous cognitive processing (Perry et al., 2012). Within dynamic environments, the diagnostic value of information will fluctuate, depending upon the nature of the situation. For instance, a rapid change in airspeed is a highly predictive cue that pilots use to diagnose wind shear events. However, this same information holds less diagnostic value for detecting an abnormally high sink rate. Therefore, there needs to be some consideration as to how reduced processing systems will manage the fluctuating diagnostic values of information to ensure that the systems provide operators with the most critical cues associated with the situation, at the appropriate time.

Previous approaches to assessments of the diagnostic value of information in the context of decision support systems have involved placing the onus of information reduction on the user (Morrison et al., 2010; Perry et al., 2012; Wiggins & Bollwerk, 2006). In this case, the information load is reduced by users customizing the interface by selecting the information that will subsequently be available to access. The problem with requiring operators to customise the interface in this way is that it can only be effective if the operator possesses knowledge of the diagnostic validity of information, which can be difficult for all but highly experienced operators (Perry et al., 2012). Therefore, the customizable approach to minimizing information load has limited viability.

A more promising application of reduced processing interfaces lies in systems that can automate the process of information reduction by detecting and presenting the most diagnostic information relating to the situation. Through the analysis of data regarding the system state, combined with logic that defines the presentation of information, systems can be designed to adapt the display of information to the demands of the situation (Lavie & Meyer, 2010). The advantage of an adaptive approach to the application of reduced processing support systems is that the quality of the information reduction process is not dependent on the operator's knowledge of the diagnostic validity of each piece of information. Indeed, the automatic configuration of a reduced processing support system, designed to automatically display the most diagnostic information, can result in performance improvements in inexperienced operators that exceed those of a customizable display (Perry et al., 2013).

Considering the need for adaptive systems, the successful application of reduced processing support interfaces requires two key system requirements to be satisfied. First, the system must have the ability to collect and display up-to-date data regarding system parameters. Due to the dynamic nature of high-reliability situations, operators must receive current information about the state of the system to maintain an accurate mental model of the situation. The second requirement is that the system has the capacity to analyse the available data to classify the system state. The automated classification of the system state enables the application of adaptation logic that specifies the information that is appropriate to display to the user, based on the current state of the system. The capacity to adapt the display of information to the system classification can ensure that the operator is provided with information that embodies the highest diagnostic value for the situation.

The level of system sophistication required to enable the application of reduced processing support systems means that a system-driven approach is likely to yield the greatest potential in supporting performance in high-technology environments such as aviation. As an example, consider the emergency warning systems in modern commercial aircraft. Aircraft warning systems can detect potential emergency events through the analysis of data regarding key aircraft parameters. Following the detection of an emergency situation, warnings can be provided to the pilots that relate specifically to the current event, providing the capacity to issue warnings directed towards a range of different events, including an approach to aerodynamic stall situations, conflicting air traffic, and/or the presence of wind shear.

Once the system state has been automatically classified, the information display can be adapted to facilitate operators' processing of the most diagnostic cues. The implication is that reduced processing diagnostic systems can be engaged to assist the initial diagnosis of the event, particularly where the event has a sudden onset and invokes an emotional response. However, one of the concerns with automated systems has been a reduced capacity for human operators to intervene due to reduced situation awareness (Kaber & Endsley, 1997). Reduced processing diagnostic systems are intended to reduce the cognitive load associated with diagnosing the system state, thereby potentially reducing the impediments to operators to actively engaging the control loop.

Continuing with the example of aircraft warning systems, Figure 11.1 illustrates how a reduced processing interface might have displayed information relating to the stall situation in AF447.

The aim of the diagnostic system is to integrate the information relating to the emergency event into a single display that will facilitate an evaluation of the validity of the warning. For the purpose of this example, three key stall cues have been

STALL	
ANGLE OF ATTACK	10⁰ ↑
AIRSPEED	200 KTS ↓
VERTICAL SPEED	2500 FPM ↑
ACTION	NOSE DOWN

Figure 11.1   Example of a reduced processing diagnostic display relating to the stall event in AF447

selected, the aircraft's angle of attack, airspeed, and vertical speed. In conjunction with the system classifying the stall situation and issuing the associated warning, the values of the three key stall cues could appear on an integrated display to provide pilots with the information needed to evaluate the validity of the warning. The arrows next to each feature value portray trend information which is intended to facilitate an awareness of the dynamic change in the three key indicators.

In theory, a reduced processing diagnostic interface should enable the accurate interpretation of the validity of the stall warning. The display directs attention to the key indicators associated with the stall event detected by the system. The elimination of irrelevant information would likely facilitate the pilots' ability to draw the necessary associations between the three displayed indicators. The proximity of the indicators reduces the need for information to be held in working memory, thereby enabling the temporal integration of task-related information (Wickens & Carswell, 1995). Following the activation of the stall warning, consultation of the diagnostic system would enable pilots to establish that both the angle of attack and vertical speed are high and increasing, while airspeed is falling. These are conditions consistent with an approach to stall.

A display that facilitates the interpretation of the indicators associated with an emergency situation would likely aid the verification of the legitimacy of an issued warning, since more than one source of information is presented. Following the confirmation that an emergency warning is valid, recovery from the situation is reliant upon an operator's ability to identify and execute the appropriate corrective actions. To reduce the potential for performance to be constrained by the failure to recall an appropriate procedure from memory, the diagnostic system could also be designed to advise an appropriate course of action.

The examplar diagnostic support system is not necessarily intended as a complete solution to issues that led to the accident of AF447. Rather, the accident provides a context for demonstrating how a reduced processing support system might be applied. The application of reduced processing systems is not limited to supporting diagnosis during warning events. The principles of reduced processing support systems can be potentially applied to all manners of diagnosis, provided that the system can play a role in classifying the system state and adapt the information displayed to ensure that a limited number of diagnostic cues is presented to the operator.

Conclusion

Efficient information processing is vital considering the increasingly complex systems that humans are required to manage and operate. Operators who have not acquired sufficient experience to have developed efficient processing strategies suffer an increased potential that their performance will be degraded due to cognitive overload. However, systems that are designed to reduce superfluous cognitive processing hold considerable promise for managing the performance of less-experienced operators and protecting system integrity, particularly during skill acquisition.

Further research is required to develop the confidence that the application of reduced processing diagnostic systems will, indeed, aid the performance of less-experienced operators. There is a potential risk that the inclusion of additional information displays could have the unintended negative consequence of increasing, rather than decreasing, cognitive load. Therefore, research is needed into the application of reduced processing systems for dynamic diagnostic tasks to determine whether the processing benefits that have been found on static tasks are maintained in situations that are more complex and dynamic.

Chapter 12

Diagnosis and Culture in Safety Critical Environments

Christine Owen

In safety critical and dynamic environments, performance relies on team members sharing what they know so that collective diagnoses of changing conditions can be made and decisions can be coordinated. Such environments include aviation, military operations, nuclear or chemical plant management, emergency health, and emergency management. In these domains, safety is accomplished, in part, because the meaning and significance of changing cues can be used as a collective cognitive resource (Burke, Stagl, Salas, Pierce, & Kendall, 2006; Flin, O'Connor, & Crichton, 2008). However, individual and collective diagnoses of cues take place in a social context and are influenced by the cultural norms of the group. As argued later in this chapter, there are many reasons why team members respond differently in collective contexts. In short, while individual diagnoses of changing cues are important – they are not sufficient for optimum team performance.

From this perspective, the difference between cognitive information processing approaches to expertise and socio-cultural approaches is that the focus of attention switches from the individual to the group or a community of practice. In this way, Konkola, Toumi-Grohn, Lambert, and Ludvigsen (2007) argue that expertise is not just developed inside the practitioner's head but also 'expands the structures of knowledge to include not just mental and symbolic representations but also … recurring patterns of social practice' (pp. 213–14). From this point of view, what is regarded as 'good' practice is not just determined by some external normative set of regulations, but also by what the work group informally regards as important (or as unimportant) and culturally valued (or not valued).

Despite its importance, the cultural context is often overlooked and seen as some benign ambient background. However, human factors scholars and practitioners ignore the role of socio-cultural influences at their peril. This chapter sets out some socio-cultural factors to show how they influence the degree to which individual diagnosis of cues are shared. The examples provided come from two safety-critical domains – air traffic control and emergency management.

It is important at the outset to be clear about what this chapter is and is not. This chapter uses exemplars to discuss how certain socio-cultural factors enhance or inhibit team communication. This is NOT to suggest that all air traffic controllers or emergency managers operate in the way described in the selected interview quotes. To begin, it is useful to establish the role of collective diagnosis in dynamic environments and how team communication and culture interact.

Team Communication and Culture

Organizations deliver highly reliable performances when members have the ability to prevent and manage mishaps before they spread throughout the system (Barton & Sutcliffe, 2009). This occurs when team members engage in social mechanisms for monitoring and reporting small or weak signals (i.e., that something might be wrong, including with their own capacity). This shared situational awareness enables team members to adjust to these changing conditions. Members have both the flexibility required and the capability to respond in real-time, reorganizing resources and actions as necessary.

From this perspective, effective diagnosis and safety is achieved through human processes and relationships (Barton & Sutcliffe, 2009). Members provide and seek information to make meaning of the cues that they consider important (which include uncertainty and discomfort) and the team adjusts, tweaks, and adapts to this information flow which, if not forthcoming, could result in larger problems and potential failures.

Distinguishing between Professional and Situational Expertise

Barton and Sutcliffe (2009) interviewed 28 experienced firefighters and, from those interviews, extracted 62 cases of incidents that had either gone well or had resulted in poor outcomes. A key difference between those that ended poorly and those that did not was the extent to which individuals voiced their concerns about early warning signs.

According to Barton and Sutcliffe (2009), an important factor inhibiting the voicing of cues of concern is deference to professional expertise. This transpires in part because perceived competence and experience is spuriously conflated with situational expertise. While leaders may have considerable expertise based on many years of previous experience, in dynamic situations, this does not necessarily make them situational experts who see the same cues or understand them as other members of the team understand them. Where conditions are rapidly changing and ambiguous, constant updates and adjustments are needed. Under these conditions, the team needs to support and maintain collective situational expertise. Barton and Sutcliffe argue that this comes from personnel drawing on different perspectives and insights to contribute to situational expertise. Team leaders also need to be able to diagnose the situational expertise available in the communication patterns. This includes the proactive pooling of ideas and what is known/unknown and what assumptions need to be tested – key attributes of expert teams (Salas, Rosen, Burke, Goodwin, & Fiore, 2006). It also involves seeking discontinuity and in making this discontinuity visible to others where needed, particularly in complex and sometimes novel events.

Situational expertise is likely to become more important given the increasing complexity and abstraction found within contemporary workplaces, where, according to Engeström (2004), a new interpretation of expertise is required. He is particularly concerned with domains engaged in significant change, where little may be known about the problem(s) in need of resolution.

Barriers to Team Communication

Despite the importance of team communication and information sharing, there are many challenges that need to be overcome. First, as Wiggins (Chapter 1 of this volume) suggests, accurate diagnosis is dependent upon a repertoire of experiences and it may be that the expertise of some team members is not so well developed. Novices, for example, may simply miss important cues that signal a need to review actions. Other research has indicated that social influences play a key role.

The stakes are understandably high when human lives (including the lives of colleagues) and material assets are under threat. Risk aversion can inhibit open information sharing and speculation, particularly under conditions of chaos and uncertainty.

In a recent study, Lewis, Hall and Black (2011) conducted in-depth interviews with 36 wild-land firefighters in the United States to explore the reasons why firefighters do or do not voice their concerns, even when such concerns are recognized. While they found that some novices simply did not recognise external cues, they concluded that not speaking up was more often due to social influences that inhibit people from doing so. These social influences were linked with the career stage of firefighters. Rookies (novice firefighters) fear that no one will listen to them. Veteran firefighters – both mid-career and experienced personnel – feel more confident in raising concerns. However, they still face career pressures to remain silent. Expert veterans face fewer social pressures, but are frequently placed in demanding positions that can become distracting, or in insufficiently demanding positions where their experience leads to complacency with its associated attentional problems.

There is also a wide body of literature that suggests that status differences lead to counterproductive hierarchical communication patterns that negatively impact safety-critical performance (Lewis et al., 2011; Edmondson, 2003; Nembhard & Edmondson, 2006; Barton & Sutcliffe, 2009). In the health industry, Edmondson (2003) found that, although nurses witness and experience a variety of problems and employ a number of creative solutions to resolve emergency issues, they generally did not communicate these issues to others in the medical hierarchy. For example, in the case of medical malpractice, physicians (the high-status members of a team) were reported to have ignored important information communicated by nurses (low-status members) and nurses also withheld relevant information for diagnosis and treatment from physicians.

Organizational Culture and Expertise

Organizational cultures are defined as the 'habits, folkways and norms that shape action' (Westrum, 1993, p. 401) and the 'set of understandings or meanings shared by a group of people' (Louis, 1986, p. 74). Cultures form within and between groups in work organizations because individual and collective experiences are concentrated in work tasks and roles in particular ways. The focus here is on a 'bottom up' view, where organizational culture is shaped both by the meaning-making and interpretations that individuals give to their workplace experiences, and where those interpretations become collectively shared.

It is contended that workplace cultures are revealed in the way that people communicate their understanding about their work (Alasuutari, 1995; Schein, 1996; Scheeres & Rhodes, 2006); their shared (implicit) norms of behaving (Balthazard, Cooke, & Potter, 2006; Augustinos, Walker, & Donaghue, 1995); the informal language that they use (Smiricich, 1983); the stories they tell (Trice & Beyer, 1984; Murphy & Hall, 2008); and the stereotypes they employ to account for in-group and out-group membership (Fine, 1996; Holland, Lachicotte, Skinner, & Cain, 1998; Lawrence, 2006). Cultures tie the actions of individuals to a particular group (or groups) and reveal, through justifications for group membership, what is collectively valued within the group (and what is not) (Fine, 1996).

Cultural norms of behaviour can also exert conformity and, hence, deter learning because they focus attention on valuing certain kinds of practices (and not others) and because they can also inhibit the questioning of those practices. Weick (2001) called this 'socially organized forgetting'. The identification with particular work-group norms contributes to individuals' social or occupational identity. Moreover, heterogeneous work organizations may exhibit cultures that are multiple, overlapping, or contested. Because of this heterogeneity of group membership in organizations, cultures can consist of shared, partly shared, non-shared, and/or contested values, beliefs, and norms.

In the United States, parochial turf boundaries may have increased the number of lives lost during the World Trade Center attack in September 2001. Marcus, Dorn, and Henderson (2006) noted that historic rivalries between the New York Police Department and the New York Fire Department inhibited information-sharing between the two groups, with drastic consequences. The police, from the vantage point of their helicopters, could see that the towers were about to collapse and conveyed this to their people on the ground, but this information was not shared with the firefighters who continued to stream into the building. In Australia, cultural impediments to information-sharing between two incident management teams of different organizations were held to be partly responsible for the tragic deaths of many in one of the more devastating 2009 Victorian bushfires (Victorian Bushfires Royal Commission, 2010).

The importance of socio-cultural context and organizational culture is illustrated in the following empirical work in which socio-cultural factors are shown to both constrain and enable team communication and the development of situational expertise.

The Empirical Studies

Interviews and observations were undertaken in two safety-critical domains: air traffic control and emergency management teamwork. In air traffic control, the author examined the implementation of teams and other technological and structural changes over a 10-year period in Australia and, subsequently, has continued to work in the industry (see for example, Owen, 2001, 2005, 2008, 2009a, 2009b; Owen & Page, 2010). More recently, the author has been engaged in various research projects in the emergency management industry and has led teams examining emergency management teamwork in emergency operations centres in Australia as part of the work of the Bushfire Co-operative Research Centre (see also Curnin & Owen, 2013; Hamra, Hossain, Owen, & Abbasi, 2012; Owen, 2014a). The results and discussion presented in this chapter afford an opportunity to look across these two datasets and to consider the following question: *What are the socio-cultural elements that enable and constrain communication patterns that support collective diagnosis in teams?*

Research Methods Employed

The experience of teamwork in both of these domains was investigated first through semi-structured interviews and then through observations. A semi-structured interview script was developed based on issues around teamwork communication that have been identified in the literature. In the interviews, an experienced qualitative researcher used open-ended probing questions to elicit participants' perceptions of what enabled and constrained teamwork and performance in their respective domains. Recordings were transcribed verbatim immediately following each interview, with any identifying data removed.

In the air traffic control sample, 61 interviews were conducted with air traffic control instructors (n=25) and trainees (n=36). These interviews occurred with personnel involved in the main air traffic control functions (approach: n=13; arrivals: n=17; enroute: n=21; and tower: n=10) across three centres in Australia. In the emergency management teamwork sample, 50 interviews were conducted with emergency management team leaders across five Australian jurisdictions (Queensland, New South Wales, Tasmania, ACT, South Australia). The air traffic control instructors interviewed had experience ranging from two to 25 years. All of the emergency managers interviewed had over 15 years of experience in the fire and emergency services industry and a median 11 years of experience working in emergency management teams. In addition, observations were conducted in both domains. In air traffic control, the observations occurred as part of the fieldwork. The method developed in the aviation industry led to observing emergency management teams and their leaders using video and audio recordings in both team exercise simulations and real-time emergency events (see Owen, 2014b). Prior to each study, institutional ethics approval was received and informed consent was obtained from all participants.

In both studies, the researcher adhered to strict qualitative research principles (Denzin & Lincoln, 2004; Guba, 1981), which supports the rigour of the work.

Findings

As mentioned earlier, the histories of collective experience are the breeding ground for culture. It is perhaps not surprising that both domains (air traffic control and emergency management) emerged out of the military. In the case of air traffic control, civilian aviation in Australia was born following the Second World War, when returning servicemen populated both aircrew and ground control. So too, firefighting and emergency services organizations have, historically, had strong ties with the military and many started out as part of Australia's Civil Defence. This historical legacy is still present in, for example, the ranking structure (e.g., captains, commanders) used within the emergency services.

In both domains, pushing the limits and managing risk has also been part of cultural history. Indeed, the aviation industry was born from the risk-taking of individuals attempting to do something that was previously not possible. The stereotypical image of success within the aviation industry is characterized by the aircraft pilot: 'a single, stalwart individual, white scarf trailing, braving the elements in an open cockpit' (Helmreich & Foushee, 1993, p. 4). These stereotypes led to acceptance, indeed celebration, of norms associated with 'independence, machismo, bravery, and calmness under stress' (Helmreich & Foushee, 1993, p. 4).

These histories support what has been referred to as a 'can-do' culture in both domains. This culture values high levels of performance and sometimes tacitly supports risk-taking and pushing safety to the limits. Heroic stories about legendary risks taken in firefighting, for example, continue to pervade the media.[1]

Both industries also contain a high proportion of men, and cultures of masculinity have been noted (Owen, 2013; Tyler & Fairbrother, 2013). Hegemonic

---

1   The thin brown line: The day the RFS stood between the Blue Mountains and Armageddon, *The Australian*. Available at: http://www.theaustralian.com.au/news/features /the-rfs-stood-between-the-blue-mountains-and-armageddon/story-e6frg8h6–1226781 702099?sv=41acb2f4582e574612f4d41f33b1dfb6#mm-premium [accessed 15 December 2013].

masculinity[2] is reproduced and reinforced through media and through social interaction. In the media, men's performance is glorified as heroic, through their physicality, their daring behaviour, their power, and their emotional detachment. Lois (2001) notes that masculinity, but not femininity, must be constantly proven. It is also important to point out that norms of behaviour regarded as appropriate within a masculinist culture can be practised by both men and women. In both domains, interviewees told stories of how important it was to be able to perform. A participant working in air traffic control recorded that:

> I think the sort of people that we have, like myself, was we wouldn't say 'No'.
> We would keep taking aeroplanes on because, after a while, it is like a big sort
> of caramel lolly or something – you can't break it up into small pieces, you just
> keep going and keep going and keep going. You can't bite it off or say 'Stop'.

In the fire and emergency services industry, the value on performance is also explained as:

> People will always be people and I don't mean to be sexist but you know, boys
> will be boys and the testosterone gets flowing and boys and are very competitive
> or want to prove a point ... and there's always you know, the tribal instinct
> coming out in all of us [being stand-offish], we're only being human.

Good performance when controlling air traffic is what many controllers value in both themselves and in others. It enables those with 'The Right Stuff' (i.e., the ability to perform) to show their value to others in the workspace. In safety-critical organizations, understanding the theories-in-use by workers as they go about their daily tasks provides important insights. Drach-Zahavy and Somech (2011) take a socio-cultural approach to explore nurses' implicit theories and their heuristics in decision-making about when and how to follow safety policies (or ignore them). They suggested that while non-adherence to safety rules might seem random, it is, in fact, systematic and predictable. They show how a reluctance to comply with safety procedures is not due to ignorance but instead, derives from a sense of professionalism about their ability to care for patients even at the risk of unfavourable consequences. One heuristic that is pertinent in this context is that 'professionals do not seek help'. According to Drach-Zahavy and Somech (2011), nurses need to present themselves as self-reliant and fully capable of working independently without assistance. It can also be argued that these deviations become part of a cultural norm as to how work is undertaken within the nursing unit and, therefore, becomes part of the culture of this team.

---

2   Masculinity (Kimmel, 2008) is seen as hegemonic in that it is dominant over alternative masculinities held by, for example, gay men or nurturing fathers (Kimmel, 2008; Connell, 1994).

The 'professionals do not seek help' heuristic was observed in both air traffic control and in emergency management. In the following transcript, the controller inhibits the possibilities of increasing his own options for handling a problem by refraining from asking a team member for help, because the norm of practice within his team does not support such inquiry.

Interviewer:
Under what conditions would you ask?

Respondent:
There have been times in the past when I've been training or I've had a question or something I couldn't quite understand and I think 'Would he know? I'll ask'. But 'No, I'll pick it up'. And sometimes you do [pick it up] and sometimes you don't, until later [and] you think, 'oh God! Is that what they meant?' And it can be little things. ... [But] 'I don't want to look stupid, I should know that' ... 'Oh well, I'll pick it up'. I'd hoped!

For this controller, the process of seeking information from another was hindered by his reluctance to reveal his own lack of knowledge. In this case, generating alternative courses of action was constrained because of a belief held by the controller that 'he should know'.

The Cultural Performance of Taking Control

Personnel in high-risk or high-consequence environments, such as air traffic controllers and emergency managers, share similar cultures that also support a particular 'presence'. Goffman (1960) regarded behaviour in everyday life as a performance, with many similarities to theatrical performances. In Goffman's (1960) terms, the main objective is to sustain a particular definition of the situation and to behave in a certain way that makes an implicit statement about what is real and important in this interaction.

Impression management is a cultural accomplishment prevalent in many professions. In law and medicine, for example, a particular type of 'gravitas' is needed. Health professionals require a particular type of demeanour that must be learned if they are to earn the trust of patients, including touching the skin of others in a way that is professional in its inquiry and not intimate. The cultural accomplishment of impression management is also important in air traffic control and in emergency management.

Good controllers exercise good 'judgement' and have confidence in the decisions they have made. These collectively held values and beliefs are salient because they play an important functional role in the smooth operation of the air traffic system, as the following controller explains:

> You've got to be confident in your memory, what you remember, how you
> approach things. You've got to achieve the pilot's confidence straight away. He
> doesn't give you a second chance. If you do something and he feels ... like ...
> He's ... putting his life in your hands, and if your confidence isn't there, or he
> doesn't feel it's there, then he's going to be nervous. And that's going to make
> things more difficult, because he's going to want to know more information,
> then you're going to get further behind, and it steam-rolls itself.

Controllers must display confidence in their own decisions so that they can move
on to the next problem and they must communicate confidence to others. The
impression of confidence needs to be conveyed to gain the confidence of others.
Establishing an impression about one's confidence thus becomes part of one's
self, as projected to others in the aviation system. However, with the cultivation
of confidence must also come a cultivation of doubt, as controllers need to
maintain vigilance against what they call a 'fat, dumb and happy' attitude of
complacency. Therefore, there is a need to maintain a balance of self-monitoring
between overconfidence and cautiousness. Controllers need to be both confident
in their decisions as well as constantly check and scan the situation to identify
unanticipated disturbances.

Although displaying confidence and being 'in control' is absolutely necessary
to be able to undertake safety-critical work in a dynamic environment, this display
becomes problematic when individuals begin exuding confidence because they have
invested in a social identity. This kind of display shuts down team communication
practices. During the data collection within air traffic control, there was a serious
near-miss incident where a controller failed to heed the warnings of others near
him of a potential breach of airspace separation:

> There was an incident recently you've probably heard about. And the controller –
> he had that problem. He was quite a good controller. / He had a problem, in the
> fact that he felt he had to prove he was a good controller every day/ And he
> wouldn't admit that he was wrong – until it was so late, and he'd been warned
> by so many people, that he was wrong.

In accounting for why the controller mentioned in this example ignored the advice
and warnings of those nearby, the following member of a team working the same
sector observed:

> Controller:
> But this [display of confidence] had been worked up over a number of years. He
> wasn't like that when he first got a rating. That had been built up.
>
> Interviewer:
> So did the environment help build that up?

Controller:
I think the environment promoted him to build that up, particularly his time at [name of Centre]. I saw it [that way of working] in [name of Centre] until my eyes bled!

Interviewer:
You saw what?

Controller:
'I am a super controller' You know? *'keep sending me aircraft'* I saw and heard this every day. And nobody trusted anybody else.

Managers are sometimes caught in 'webs of significance of their own making' (Munro, 1999, p. 635). In emergency management, presence is also demonstrated by exhibiting behaviours characterized by being calm and in control. This is especially problematic if personnel believe they must continually prove their abilities and this involves displays of confidence or presence. In the interviews within the emergency management domain, participants were asked to comment on the factors that impede them gaining an awareness of the emergency situation, a precursor to effective action.

The most frequently reported theme was a rigid or autocratic management style, or as one participant describes, it is 'a 'my way or the highway' sort of attitude'. Another theme related to personal attributes that inhibited the sharing of information. As one participant explained, this was represented by 'bravado and a lack of respect for others'.

In emergency management, this portrayal of performing to provide a particular impression is also linked to a need to be seen to be in command, which can lead to autocratic communication styles. In another comment, an interviewee discusses the impact of what he calls the 'command and control type attitude' and its negative impacts on communication and cooperation:

You can see it all the time. An effective officer builds a really quick relationship with their counterpart and explains in terms they can understand and creates a rapport with them and things work. Other people adopt this really command and control type attitude that 'you can't come in here 'cause this area is mine' and it just sets this chain of like interpersonal conflict that puts everyone at risk.

While establishing control is clearly part of a leader's responsibilities, it is important that this not be conflated with a style of communication that does not actively support the collective development of situational expertise.

Not all practices operate to constrain communication. Some operate to facilitate experts to become conscious of what situational cues they are detecting and to share these with newcomers so that they may learn.

Overcoming the Limits of Professional Expertise

In the air traffic control study, an unintended consequence was observed after the implementation of a structural industrial relations change, which was initiated to support capacity-building within the organization. The change was implemented as part of the enterprise bargaining agreement and required all fully rated controllers to perform the role of instructor and mentor for new trainees. While intended to support trainees, the change had an unintended effect on the expertise of the instructors (Owen, 2009b).

The process of instructing sets up opportunities for controllers-as-instructors to observe and reflect on the job of controlling – something that is difficult to do when engaged in temporally and cognitively demanding work that cannot be stopped. When controllers instruct trainees, they usually sit or stand behind the trainees and watch them work at the console. The practice of watching another do the work of controlling enables the controller to view the work activity from a different perspective. Controllers commented that having to become an instructor forced them to re-examine their own knowledge base. Stepping back from the job at the console had the advantage of giving the controller a different (wider) perspective as explained by the following interviewee:

> In some ways, it improves you because you become more aware of what is going on around you and you look at other ways of doing things and being forced to sit back and watch it in detail – you analyse it a lot more. There is an old saying on Approach 'he's been doing the job for a year now; it is time for a trainee so he can really look at the whole thing'. Quite often, training officers are selected from the point of view to improve the training officer not for the trainee's benefit. There is quite a bit of that going on, and there's a lot of truth in it.

As Wiggins pointed out in Chapter 1 of this volume, expertise is predicated on a highly evolved knowledge base as well as highly automated skills that are developed over years of practice. In these cases, individuals who have been performing the task over a long period of time forget which maxims and rules they are invoking in undertaking the work, making the skills learned opaque, even to themselves. The requirement to teach another pushes experts beyond their automated task performance towards an explicitly conscious awareness of that performance. This is of particular benefit because 'explaining one's thinking to another leads to deeper cognitive processing' (Bereiter & Scardamalia, 1989, p. 362). Team leaders also play a critical part in establishing the communications climate within the team.

Diagnosing Situational Expertise: Riding, Spanning, and Crossing the Boundaries

The observations conducted of emergency management team leaders revealed that more effective team leaders engaged in three team coaching roles to facilitate communication and coordination within and between teams (Owen, 2014b). Team leader feedback was directed towards coaching team members to address three broad purposes:

- addressing the temporal demands of the task and making adjustments needed in response to changing conditions (i.e., riding the boundaries of the team);
- integrating their activities within the team (i.e., spanning the boundaries within the team); and
- reporting and seeking cooperation with others outside the team (i.e., crossing the boundaries between teams).

Boundary Riding

Coaching for 'Boundary Riding' included activities that were designed to assist team members clarify and exercise their own responsibilities and to be mindful of the impact of those responsibilities on others. This also involved coaching team members in their responsibilities to contribute to team thinking and problem solving. The emphasis was on constantly calibrating and adjusting to the temporal demands in managing the event and sharing information about any perceived changes.

Boundary Spanning

The intention of 'Boundary Spanning' is to ensure that there is effective integration within and between functional units within the team. In so doing, there was an emphasis on updating inter-positional knowledge. Having inter-positional knowledge (i.e., knowledge about the environment, tasks, roles and appropriate behavioural responses required of team mates in various situations) assists with the development of shared mental models (Burke et al., 2006). Coaching for functional integration between member roles and responsibilities occurred in a number of ways. These included orchestrating to ensure that roles were linked, and engaging in blocking activity that was designed to re-route information that needed to be given to someone else. They also engaged in visible monitoring – which involved making their observations visible to others – and more actively offered and sought assistance from others, thereby influencing the climate of the team.

Boundary Crossing

In 'Boundary Crossing' the focus is on meeting the information needs and concerns of other teams operating within the emergency management structure and of external stakeholders involved in the operation. Effective team leaders accomplished this by working to align expectancies between the team and other teams or stakeholders. This involved aligning expectations in relation to both information-seeking and information-giving between teams and/or organizations; strengthening assertiveness, which involved coaching team members to be outwardly focused; and being clear in managing external stakeholder expectations.

Conclusion

This chapter has argued that communication practices within teams are critically important but that cultural norms sometimes inhibit communication flow during diagnosis. This is dangerous in safety-critical domains, particularly where the external conditions are dynamic. Individuals may notice cues but not fully understand their meaning. Even if they do, they may not share this diagnosis with others for fear of negative consequences. Team leaders, human factors change interventionists, and training instructors can use indicators of cultural norms within groups to undertake their own diagnosis of the communication patterns in a group. Therefore, human factors practitioners can engage in a useful analysis to evaluate the degree to which socio-cultural and structural features of work organization act as enablers or inhibitors of team communication and the development of situational expertise. If the domain is one where sharing information to support collective diagnosis and the development of collective situational expertise is important, then a number of interventions are possible.

Workplace designers can examine how structural and cultural processes enable and constrain sharing information about the meaning of cues. Structures such as making all competent members responsible for the learning of others is one way in which experts can recover their professional expertise, by bringing back into consciousness their own strategies for identifying and addressing important cues.

It is also desirable to confront cultural norms of 'individual-performance-at-all-costs' because of their potential negative consequences. Team leaders need the skills to be able to undertake their own diagnosis of the communications climate and the degree to which team members seek to provide information to contribute to team situational awareness and, therefore, collective situational expertise.

The argument presented here also supports the contention that we need to rethink how expertise develops. In dynamic environments, professional expertise is no longer sufficient, particularly when problems include novel elements. There is a need to support the development of situational expertise, especially when team members may hold different understandings and perspectives.

Where conditions are rapidly changing and ambiguous, constant updates and adjustments are required. These adjustments can only be made if personnel are drawing on different perspectives and insights to contribute to situational expertise. Situational expertise includes the proactive pooling of ideas, what is known/ unknown and what assumptions need to be tested – attributes of expert teams (Salas et al., 2006). It also involves seeking discontinuity and making this visible to others where needed, particularly in complex and sometimes novel events.

Socio-cultural theories offer useful insights into understanding performance in safety-critical workplaces. Once socio-cultural elements are identified, it is possible for human factors practitioners to analyse existing forms of individual and collective work practice to evaluate the ways in which what has been learned supports or is at odds with desired organizational change.

Chapter 13

Diagnosis in Operations Control

Peter Bruce

This chapter explores the role of diagnosis in Airline Operations Control. It describes the airline operations environment and, in particular, the role and scope of the Operations Control Centre (OCC), which manages day-to-day operational decision-making. Attention is then given to the influence of situational awareness and expertise on decision-making processes within the OCC.

Aviation Environment

Aviation is an industry that is highly complex, dynamic, and uncertain. Safety is paramount in all facets of this industry, including the key operational areas such as piloting, engineering, and Air Traffic Management (ATM). Indeed, a safety mindset is fundamental to all who work in and around the aviation industry, because the consequences of errors can be extremely serious (Woods, 1988). Consequently, safety is rigidly maintained by a comprehensive regulatory framework of standards and practices.

With safety always in mind, a commercial airline also operates within a number of critical economic margins, necessitating both the minimization of costs (Lederer & Nambimadom, 1998) and the most efficient and productive use of assets. This focus has become increasingly important in recent years as competition among airlines intensifies. An airline operates expensive aircraft, consumes vast quantities of fuel that is subject to considerable cost volatility, employs highly skilled personnel, and must meet rigid regulatory policies and standards worldwide. Further, it has to perform, both commercially and operationally, to required standards to satisfy a number of stakeholders, including passengers, business and alliance partners, shareholders, and government. As a result, the airline's operational performance is closely scrutinized and measured. For example, operational performance is measured in terms of the status of its on-time performance, the number of flight additions and cancellations, operating revenue against network load factors (the ratio of occupied seats to offered seats), together with other parameters. Throw a highly and increasingly competitive environment into the mix and it is not difficult to appreciate some of the pressures under which an airline has to operate.

A key characteristic of the aviation industry is the nature of its unpredictability (Wiggins & Stevens, 1999), meaning that an airline is subject to considerable disruptive influences. Many observers will be familiar with the lengthy flight delays or cancellations that result from any combination of factors influencing an airline's operations on any particular day. The airline's performance is often measured by its response to such disruptions, in terms of how well it is able to match the passenger's travel requirements with the stated schedule. If the airline's response is slow, upsets or changes passengers' travel plans considerably, or isn't communicated appropriately, customer loyalty is likely to be compromized. Therefore, in responding, the airline must satisfy a number of objectives. Naturally, the prime motive must be to provide a satisfactory remedy that mitigates the disruption to its passengers. However, in doing so, the airline must also take into account the effects of internal disruptions to key areas, including pilot and flight attendant crewing and maintenance schedules. In addition, many other operational, commercial, and associated support areas are affected by disruptions in one way or another. Therefore, to achieve the desired response, there is often a need for rapid intervention, requiring collaborative decision-making by key departments and personnel, culminating in the formulation of well-thought-out and readily executed plans. In these situations, decisions yielding optimal results are crucial, while the consequences of poor decision outcomes may be commercially and/or operationally disastrous.

Introduction to the OCC

The management of the airline's network of schedules on a daily basis is at the core of the airline's operations. Within this core lies the airline's 'nerve' centre, referred to by a variety of names, including the Airline Operations Control Centre (AOCC), Integrated Operations Control Centre (IOC), or System Operations Control Centre (SOCC). However, it is most commonly referred to as Operations Control Centre (OCC), and it is this term that will be used throughout the chapter.

The OCC has responsibility for the control of aircraft to ensure economical, operational, and commercial efficiency (Williams, 1967). To fulfil this responsibility are a group of expert decision-makers who coordinate and control aircraft movements, often referred to as Movement Controllers. In recent years, it has become increasingly common to physically integrate the core operations control structure with representatives from pilot and cabin attendant crewing, engineering, flight dispatch, commercial and customer service functions, air traffic control liaison, airport liaison, and meteorology. However, even in the absence of such integration, key areas normally lie in close geographic proximity to the OCC. Most airlines locate the centre at a major hub or their own headquarters, where access to, and by, senior management, and other key stakeholders, is readily available.

The OCC is provided with a planned schedule of flights, reflecting the expectations of booked passengers and setting the parameters within which airline operations are based. The schedule indicates the locations to which the airline operates, the times at which the flights are planned to operate (Bazargan-Lari, 2004), and the specific aircraft types to operate those flights. The plan for the operating day also takes into account maintenance requirements, such as aircraft servicing or other work to be carried out during the forthcoming day(s) or night(s). Finally, the plan ensures that all flights have the requisite complement of flight and cabin crews.

The primary scope of responsibility of an OCC is to exercise direct operational control of flights within a particular period of operation. For domestic operations, for which most flights typically begin and conclude within a calendar day, the emphasis is on that day's operation. However, for international operations, the focus may be over several concurrent days. The chief function of an OCC is to ensure that, as far as possible, operations mirror the planned schedule. This is achieved by monitoring the progress of flights, identifying potential or actual operating problems, and by taking corrective actions in response to disruptions (Kohl, Larsen, Larsen, Ross, & Tiourine, 2007).

There is great emphasis on maintaining schedule integrity in airline operations. However, despite meticulously planned schedules and appropriate resourcing, airlines constantly face a multitude of operational disruptions that stem from operating in such a highly dynamic and uncertain environment. Perhaps the most prominent of these disruptions is adverse weather conditions including fog, thunderstorms, ice, snow, typhoons, floods, crosswinds, headwinds, and extreme temperatures, all of which may impact various phases of flight and, therefore, the planned schedule. Maintenance or technical problems may also cause disruption through unserviceability in-flight or between flights. In addition, there may be performance limitations or operational restrictions in the utilization of the fleet due to servicing or other requirements. Delays attributable to crewing may occur as a result of crew rest requirements, sickness, positioning for duty, and/or restrictions on flying hours or total duty hours. Besides these more frequent and familiar events, there are many other occurrences that disrupt flights. For example, airline-accountable delays could be the result of ramp, baggage, catering, industrial, check-in, or other problems. Increasingly of concern are external causes such as Air Traffic Control (ATC) requirements and airport-influenced delays (e.g. congestion both in the air and on the ground, operational limitations, power supply problems, and runway and taxiway configurations).

Managing Operational Disruptions

Managing these disruptions in the OCC can be extremely challenging. Seldom do isolated operational problems occur that are readily fixable and that cause minimal disruption to the schedule. Subject to the type and extent of the problem, and the number of concurrent problems, OCC controllers have a number of operational actions that they can take. The simplest action may involve absorbing delays on an impending flight or pattern of flights, which may isolate the disruption to a minimum number of aircraft (and hence, passengers). Other techniques may involve swapping aircraft patterns (such that each aircraft picks up the other's flying commitments for the day) to maximise the use of buffers or long periods on the ground as a means of absorbing delays.

With the mixture of fleet types that are typically operated, the opportunity for swapping *between* types is available but, inevitably, at some further cost. The obvious adjustments in seating capacities (one aircraft may have 180 seats; the other 240 seats) may result in an overcapacity situation on one aircraft, with resultant offloads, and even the simplest of type changes have to take into consideration changes to crews (as crews cannot be endorsed concurrently to fly completely different types of aircraft), catering, fuel load, aircraft weight and load distribution, different ground handling equipment, as well as any maintenance issues.

In more complex disruptions, major upheavals may result in numerous and considerable delays, cancellations, diversions, or additional flying commitments. More often than not, a single airline's fleet size and sheer volume of scheduled activity is such that disruptions affecting solitary flights are uncommon. A weather problem such as fog enveloping an airport is quite likely to affect several of the airline's flights operating into that airport. This may result in multiple diversions to alternate airports with hundreds or thousands of passengers ending up in the wrong destination, crews out of position and out of duty time, aircraft incorrectly positioned for their next flight commitments, and/or a severely disrupted maintenance programme. The diversionary airports themselves can become severely congested, lacking the resources to manage the influx of aircraft and, of course, the impact of a significant disruption on the air traffic system exacerbates these effects. The recovery plan, in this event, is extensive and can incur significant costs to the airline and considerable disruption to its passengers.

Managing an individual disruption may already be sufficiently complex in terms of the challenges encountered, but the intensity heightens with simultaneous disruptions. For example, fog at Airport A, causing a number of aircraft to enter holding patterns and others to divert to alternate airports, is likely to upset several aircraft patterns, but may only be one of several problems occurring at the time and being managed by the OCC. In other parts of the airline's network, there may be a technical or maintenance issue, a refuellers' strike, damage to an aircraft due to a ground incident, and/or air traffic control delays.

Managing such disruptions in such an intense, dynamic environment calls for highly skilled individuals within the OCC. Primarily, the roles of these controllers include continual monitoring and assessments of the progress of some or all of the airline's fleet, involving the sourcing and interpretation of key information that may be limited, unreliable, or may change over time. In collaboration with other stakeholders, controllers must make an assessment of the situation and then formulate a series of decisions, often in short time-frames, to resolve, or at least reduce, the impact of the problem.

Decision-Making

The decision-making process has been regarded as one of the most important and pervasive of human activities (Suvachittanont, Arnott, & O'Donnell, 1994; Hogarth, 1987). It is a process involving the generation and evaluation of possible solutions to identified problems, with a view towards implementing the optimal solution. Interest in decision-making has stemmed from a desire to understand how people make decisions in a variety of situations and thereby develop ways to improve the effectiveness of decision-making and reach better outcomes. In the aviation industry, this interest has predominantly focused on commercial and military pilots and air traffic controllers. One reason for this focus has been the potential implications of decisions in terms of safety. Although less focus has been evident for decision-making in OCCs, there has been extensive interest and research in the development of automatic optimization models to provide disruption recovery strategies. However, despite this work, no universal model of optimization has yet been developed to facilitate the tasks required to solve complex operational problems. Barnhart, Belobaba and Odoni (2003) suggest that this is due to the size and complexity of the situations, and the fact that these models are designed to deal separately with schedules, fleet assignment, maintenance, and crew scheduling. Further, it appears that contemporary computer-based tools have difficulty coping with the time-constraints that are characteristic of disruption management (Clausen, Larsen, Larsen, & Rezanova, 2010). Therefore, as Lan, Clarke, and Barnhart (2006) note, the majority of airline OCCs tend to recover from disruptions through manual intervention.

Decision-Making Under Time Constraints

The impact of time constraints on human decision-making is particularly evident within OCCs, but this is largely dependent on the type of operation being controlled. Decision-making environments for international and domestic operations embody very different characteristics. In international operations, flight stages can be quite long (typically up to 18 hours), and aircraft may only operate a relatively limited number of flights per day. They also tend to be parked at airports for relatively long periods of time between flights. This accommodates cleaning, fuelling, and periods of unloading and loading. In general terms, decision-making in international operations may occur over relatively longer time-frames. Nevertheless, there are constraints to consider, including the relative infrequency of aircraft operating between the same or similar city pairs, resulting in fewer options for recovery from disruption.

By contrast, domestic environments are typically more intricate as they are characterized by a high volume of flight stages during a day (10–12 stages), they operate shorter flight stages (often between one and four hours), there are nightly maintenance requirements, complex crewing commitments, and aircraft usually remain on the ground for relatively short durations between flights (often between 30 minutes and one hour). In addition, the nature of the domestic network schedule is such that a number of flights operate within particular geographical areas and time periods. Therefore, between two ports, there may be a scheduled flight every hour or even half hour. This is a double-edged sword, simultaneously, providing more resources for controllers to juggle, while also increasing the volume of operational changes and the complexity in the event of a disruption. Operations controllers who manage domestic disruptions make a number of decisions within very short time-frames and, as a consequence, decision-making in domestic operations is far more intense than decision-making in international operations.

Previous studies have examined the influence of time constraints on decision-making (e.g. Betsch, Haberstroh, Molter, & Glöckner, 2004), with much of this effort focused on the ways that decision-makers might adapt their cognitive processing in response to the time constraints. Part of the decision-making process focuses on ascertaining appropriate information. In OCCs, information may be scarce or abundant, reliable or untrustworthy, or simply unqualified. For example, information about a technical problem in an aircraft may be relayed to the OCC by a well-meaning but ill-informed staff member who is not technically qualified nor authorized to deliver the message. The difficulty increases for OCCs during a time-constrained event, when there is a need to discern between this source and a more appropriate channel. Chu and Spires (2001) suggest that, if the amount of time available limits the amount and quality of information necessary for making effective decisions, the response of decision-makers may be to process information faster, giving rise to errors or oversights. Other suggestions have focused on the compromized decision. For example, Bronner (1982) suggests that, in situations where time for decision-making is limited, decision-makers may not be able to identify a solution to a problem. As a result, they may 'buy' time to assess the situation comprehensively and defer any decision (Cohen, Freeman, & Wolf, 1996). A negative outcome for the OCC is that controllers may only consider a limited number of alternatives in decision-making or may 'satisfice'. That is, they may select a workable alternative solution that appears to solve a problem, rather than continue to explore other alternatives that may result in a better outcome.

Situation Assessment and Decision-Making

Situation assessment is critical to being well-informed (Artman, 2000) and achieving enhanced readiness for decision-making (Kraut, Fussell, & Siegel, 2003). Its importance has been recognized in many complex, dynamic environments, including anaesthesiology (Gaba, Howard, & Small, 1995; Zhang, Drews, Westenskow, Foresti, Agutter, Bermudez, Blike, & Loeb, 2002), military flying (Matthews, Strater, & Endsley, 2004), marine navigation (Sauer, Wastell, Hockey, Crawshaw, Ishak, & Downing, 2002), driving performance (Ma & Kaber, 2007;

Stanton & Young, 2000), and medical emergency dispatch (Blandford & Wong, 2004). In these environments, situation assessment has been considered in terms of decision-makers acquiring a mental picture as a precursor to responding in a situation, and to the necessity of updating that 'picture' as circumstances change. Thus, it is evident that decision-makers in these domains must understand the need to build, develop, and then maintain a degree of situation assessment. The extent to which OCC controllers gain and maintain situation assessment appears to inform and influence their own decision-making. For example, controllers in OCCs typically use Gantt charts portraying colourful and dynamic displays to represent aircraft flight schedules over several days. These schedules are presented on one or more computer screens, subject to the extent and complexity of the airline's fleet. A time scale borders the horizontal axis, and fleet units, normally grouped by aircraft type and then by aircraft registration or tail number, are arranged on the vertical scale. A matrix of 'flight blocks' or 'puks', depicting each flight number, origin, destination, passenger loading, and many other parameters, populates the display to represent all of the flight stages that are to be operated within a time period. As flights depart and arrive, the display is updated by text, colour changes, and various warnings. Other dynamic information may also cue controllers to the status of the airline's operations at any time.

With this display in mind, OCC controllers acquire an assessment of the situation in distinct stages (Bruce, 2011). As a means of obtaining a briefing (for instance, at the commencement of a shift), controllers familiarise themselves with any current or potential disruptions to the schedule using a combination of observation and enquiry of task-related features. Through this process, they gain an initial level of awareness of current operating problems such as flight delays or existing weather situations, and of future threats, including insufficient connection times for passengers or crews, or projected performance limitations. They enhance their preliminary observations by actively seeking further information or by searching for other cues in the display (e.g. during shift handover) (Bruce, 2011).

The assessment of a situation is a precursor to situational awareness which involves the augmentation of existing information by acquiring additional, event-specific information (Endsley, 1995a). For example, a flight operating into a destination for which there was initial knowledge of a weather problem, such as fog, may prompt the controller to seek further information from a relevant expert. However, on occasion, the volume of information may become extreme. It appears that, faced with complex situations, a high volume of information may overwhelm controllers who have limited time within which to process the information. As a result, they may have difficulty coping with the volume of information, resulting in some loss of situational awareness. In OCC environments, this loss of awareness is likely to occur to a greater extent in domestic, rather than international operations, due to the reduced decision time-frames and increased activity. Therefore, while acquiring information in the familiarization stage may be important for controllers in international OCCs, it may well be critical for controllers in domestic OCCs who are likely to need a more advanced state of preparedness.

Situation Assessment and Information Completeness

The prominence of information completeness for situation assessment and decision-making has come to light in domains such as health care, medicine, driving, the military, and aviation. Of interest has been the quantity and quality of information considered for use by decision-makers (Browne & Pitts, 2004), where the quality of information refers to its usefulness, currency, and accuracy (Rieh, 2002). Ebenbach and Moore (2000) note that decision-makers need to acquire relevant and sufficient information to ensure that decision-making is based on complete, rather than partial, information, but at the same time, decision-makers need to avoid saturation. However, it is difficult both to strike the right balance and to recognise that individuals' abilities to accomplish this process differ considerably. For example, Bazerman and Chugh (2006) note the inability of many individuals to acquire the correct information at the right time. Similarly, Sarter and Woods (1991) suggest that a difficulty in acquiring situational awareness stems from the 'need for perceiving, integrating, and retrieving competing information from a variety of sources' (p. 53). In OCCs, acquiring sufficient and appropriate information for decision-making is critical due to the complexity of problems encountered in a constantly changing environment. This complexity creates a significant degree of uncertainty, which may be complicated further by both conflicting information from a number of sources, and technical information that requires expert interpretation. A further obstacle in OCCs is the sourcing of information that is not readily forthcoming. Kohl et al. (2007) point out that information relating to key operational areas such as engineering (in the case of a technical problem which may delay a flight) is typically available for the OCC, but information from a less critical area such as catering, which may equally delay a flight, is less available. Therefore, it is important for controllers to acquire precise and complete information in a timely manner, as this is likely to lead to informed decision-making.

In terms of the Gantt chart used by controllers to display their flights, the way that information is presented to controllers is important. A study by Zhang et al. (2002) of the information displays presented to anaesthetists (*N*=24) concluded that integrated information displays led to higher levels of situational awareness than was achieved when using other displays. According to Bedny, Karwowski, and Jeng (2004), the enhancement of situational awareness comes about as the operator can synthesise the information to create a holistic picture of the situation. However, acquiring too much information might lead to a cluttered display and the possible loss of situational awareness (Sauer et al., 2002). This suggests that there may be an optimum method to present flight schedules and other operational information to controllers in the context of OCCs. A conclusion drawn from these studies is that the display and quantity of appropriate information is likely to influence the way in which decision-makers undertake their assessment and thereby develop situational awareness. Insufficient or poorly accessible information may starve decision-makers of the evidence necessary to act, but an over-supply of information will serve to cloud the picture being formed in the controller's mind.

Expertise and Decision-Making

Extensive research has reported the relationship between expertise and decision-making (e.g. Chase & Simon, 1973a; Chi, Glaser, & Farr, 1988; Ericsson, 2005). This has been partly in recognition that learning from experts may lead to improvements in decision-making amongst novices. However, to date, few studies have examined the influence of expertise on the decision-making processes of OCC controllers.

Despite extensive research on the subject, identifying a robust working definition of 'expertise' has been particularly difficult because of the vast number of methodological approaches that have been undertaken. It has been variably proposed that expertise corresponds with self-acclaim, attributing an individual as an expert, with specific training, and/or the display of knowledge, expressed either in terms of performance or being consistent with other experts. None of these definitions has been readily accepted in practice. There is general agreement, however, that expertise draws heavily on experience (Chi et al., 1988; Shanteau, 1992) and there also appears to be some agreement that 10 years' experience in a specific domain is a necessary but not sufficient requirement for the acquisition of expertise within a domain (Hoffman, Shadbolt, Burton, & Klein, 1995; Simon & Chase, 1973).

Controllers in OCCs are typically drawn from other operational areas of an airline, or from other aviation organizations such as airports, air traffic control, and the military. One increasingly valuable source of controllers is the highly qualified (commercial or airline transport) pilot who usually has an in-depth knowledge of the various components of aviation and can provide a pilot's perspective of disruption management. These operators usually have a wealth of experience in a number of related domains and, typically, have acquired considerable experience within the OCC. There are also specialist roles in OCCs, with responsibility for engineering, for weather briefings and flight planning advice, for customer care, or for other operational areas. Although sometimes stationed around the periphery of the core decision-makers in the OCC, these roles provide vital information and advice to assist the decision-making process. Consequently, the cumulative experience level in OCCs is substantial, thereby contributing to the levels of expertise present.

The acquisition of expertise relies not just on gaining experience per se, but more so on the quality of that experience. According to Jensen, Chubb, Adrion-Kochan, Kirkbride, and Fisher (1995), this means having a variety of meaningful experiences, which can be achieved through exposure to novel situations and multiple types of events, rather than by the repetition of similar events (Reuber, 1997). However, acquiring expertise also requires an inherent ability, a degree of deliberate practice (Ericsson, Krampe, & Tesch-Römer, 1993) and intense preparation (Ericsson & Lehmann, 1996).

Expertise in OCCs

Experts in OCCs display a number of characteristics that differ markedly from novices. Much of this difference relates to the initial assessment or diagnosis of the situation, particularly with regards to the time spent carefully reviewing the flight display searching for actual or potential operating problems. Experts invest a relatively greater effort when commencing shift or when undergoing a thorough briefing as part of a handover (Bruce, 2011). Through this process, experts tend to develop a more complete picture of the operation, taking into account the strengths and weaknesses afforded by the schedule. Therefore, they are more efficient in identifying and responding to key cues, such as passenger loadings or extended ground times that might be associated with a disruption. Heavy loadings may well deter a controller from disrupting a specific flight if other options are available, whereas a lightly loaded flight may be targeted for cancellation. An extended ground time may facilitate an opportunity to change aircraft patterns to mitigate delays.

Experts may also actively respond to cues that are not readily apparent on the display, such as querying a change to weather conditions that may affect future operations into an airport, or ascertaining projected air traffic congestion problems. In contrast, less experienced operators may fail to grasp such comprehensive measures as part of their initial assessment of the situation. They are inclined to scan the display quite superficially, with little expectation or appreciation of problem areas, unless issues are drawn to their attention. This limited focus provides only a surface understanding of the environment and, as a result, less experienced operators are less prepared in terms of managing disruptions.

Experts and less experienced operations controllers also behave quite differently in coping with operational disruptions. The enhanced preparation amongst experts at the commencement of a shift appears to enable them to encounter a disruption with a clear understanding of the circumstances, knowledge of whom to involve (or at least from whom to seek information), and an appreciation of the consequences of alternative actions that may be considered to solve the problem.

Conclusion

In disruption management, a characteristic that sets expert operations controllers apart from less experienced operators is the ability of experts to diagnose, using cues, which options have the potential to remedy a problem and which options will not. This ability enables experts to create a range of creative and workable alternative options, and then methodically assess each of these against known or anticipated limitations (Bruce, 2011). For example, a major technical problem occurring to an aircraft is quite likely to develop into an extensive operational situation with downstream consequences for delayed flights, delayed passengers, the onward connections of those passengers, disrupted crew commitments and patterns, disruption to the maintenance program, and many other issues. In this situation, experts are likely to use a highly proactive approach to procure accurate and timely information, chiefly from maintenance and crewing areas, and rapidly determine a number of contingency steps for action (Bruce, 2011). This ability to diagnose a problem and know how to achieve one or more viable solutions is a key characteristic of experts in OCC.

Less experienced operators, by contrast, appear more haphazard in the way in which they grasp the severity of a problem. They are likely to experience difficulty drawing together the different elements because they fail to observe the 'big picture' and, thus, have difficulty assessing the consequences of possible actions. Changing circumstances, such as updated or conflicting information, cast further doubt on the ability of less experienced operators to cope with an increasingly complex scenario. Rather than demonstrating proactiveness, these operators are more inclined to wait for an event change before taking action, even to the extent of watching conditions deteriorate (Bruce, 2011).

The differences between expert and less experienced operations controllers described here highlight the frailty of human decision-making. Even with significant development in decision-aided support tools, the human factor will always be central to the processes within the OCC. Artificial tools may well bring to the decision table advanced levels of information, alternative cost analyses, and workable solutions, but the human remains an intuitive expert who can add a unique dimension to the decision-making process.

References

Abernethy, B., Neal, R.J., & Koning, P. (1994). Visual-perceptual and cognitive differences between expert, intermediate and novice snooker players. *Applied Cognitive Psychology*, 8, pp. 185–211.

Abernethy, B., Wood, J.M., & Parks, S. (1999). Can the anticipatory skills of experts be learned by novices? *Research Quarterly for Exercise and Sport*, 70, pp. 313–18.

Aggarwal, R., Mytton, O.T., Derbrew, M., Hananel, D., Heydenburg, M., Issenberg, B., MacAulay, C., Mancini, M.E., Morimoto, T., Soper, N., Ziv, A., & Reznick, R. (2010). Training and simulation for patient safety. *Quality and Safety in Health Care*, 19(Suppl 2), pp. i34–i43.

Air Accident Investigation and Aviation Safety Board. (2006). *Helios Airways Flight HCY522 Boeing 737-31S at Grammatiko, Hellas on 14 August 2005: Hellenic Republic Ministry of Transport & Communications*. Athens, Greece: Author

Alasuutari, P. (1995). *Researching Culture: Qualitative Method and Cultural Studies*. London: Sage.

Albrecht, J.E., & O'Brien, E.J. (1993). Updating a mental model: Maintaining both local and global coherence. *Journal of Experimental Psychology: Learning, Memory, and Cognition*, 19(5), pp. 1061–70.

Allison, P.D. (1992). Cultural relatedness under oblique and horizontal transmission rules. *Ethology and Sociobiology*, 13, pp. 153–69.

Altman, E.I., & Saunders, A. (1998). Credit risk measurement: Developments over the last 20 years. *Journal of Banking & Finance*, 21, pp. 1721–42.

Alvarez, G., & Coiera, E. (2006). Interdisciplinary communication: An uncharted source of medical error? *Journal of Critical Care*, 21, pp. 236–42.

Anderson, J.R. (1987). Skill acquisition: Compilation of weak-method problem-solutions. *Psychological Review*, 94, pp. 192–210.

Anderson, J.R. (1993). *Rules of the Mind*. Hillsdale, NJ: Lawrence Erlbaum.

Andersson, P. (2004). Does experience matter in lending? A process-tracing study on experienced loan officers' and novices' decision behavior. *Journal of Economic Psychology*, 25, pp. 471–92.

Ariga, A., & Lleras, A. (2011). Brief and rare mental "breaks" keep you focused: Deactivation and reactivation of task goals pre-empt vigilance decrements. *Cognition*, 118, pp. 439–43.

Artman, H. (2000). Team situation assessment and information distribution. *Ergonomics*, 43, pp. 1111–28.

Asch, S.E. (1951). Effects of group pressure upon the modification and distortion of judgements. In H. Guetzkow (ed.), *Groups, leadership, and men* (pp. 222–36). Pittsburgh, PA: Carnegie Press.

Asch, S.E. (1956). Studies of independence and conformity: I. A minority of one against a unanimous majority. *Psychological Monographs: General and Applied, 70*, 1–70.

Augustinos, M., Walker, I., & Donaghue, N. (1995). *Social Cognition*. London: Sage.

Australian Transport Safety Bureau (2007). Radiotelephony readback compliance and its relationship to surface movement control frequency congestion (Report No. B2006/0053). Canberra: Author.

Auton, J.C., Wiggins, M.W., Searle, B.J., Loveday, T., & Xu Rattanasone, N. (2013). Prosodic cues used during perceptions of nonunderstandings in radio communication. *Journal of Communication, 63*, pp. 600–16.

Aviation, Space, and Environmental Medicine, 76, B154–63.

Baber, C., & Butler, M. (2012). Expertise in crime scence examination: Comparing search strategies of expert and novice crime scence examiners in simulated crime scenes. *Human Factors, 54*, pp. 413–24.

Bacon, L.P., & Strybel, T.Z. (2013). Asssessment of the validity and intrusiveness of online-probe questions for situational awareness in a simulated air-traffic-management task with student air-traffic controllers. *Safety Science, 56*, pp. 89–95.

Baddeley, A.D. (1972). Selective attention and performance in dangerous environments. *British Journal of Psychology, 63*, pp. 537–47.

Balthazard, P.A., Cooke, R.A., & Potter, R.E. (2006). Dysfunctional culture, dysfunctional organization: Capturing the behavioral norms that form organizational culture and drive performance. *Journal of Managerial Psychology, 21*, pp. 709–32.

Bandura, A., & Locke, E.A. (2003). Negative self-efficacy and goal effects revisited. *Journal of Applied Psychology, 88*, pp. 87–99.

Barnhart, C., Belobaba, P., & Odoni, A.R. (2003). Applications of operations research in the air transport industry. *Transportation Science, 37*(4), 368–91.

Bartlett, F.C. (1932). *Remembering*. Cambridge: Cambridge University Press.

Barton, M., & Sutcliffe, K.M. (2009). Overcoming dysfunctional momentum. *Human Relations, 62*, pp. 1327–56.

Bartram, L., Ware, C., & Calvert, T. (2001). Moving icons: Detection and distraction. *Proceedings of the IFIP TC.13 International Conference on Human-Computer Interactions* (INTERACT 2001), Tokyo, Japan.

Bartram, L., Ware, C., & Calvert, T. (2003). Moticons: Detection, distraction and task. *International Journal of Human-Computer Studies, 58*, pp. 515–45.

Baumann, C., Burton, S., Elliott, G., & Kehr, H.M. (2007). Prediction of attitude and behavioural intentions in retail banking. *International Journal of Bank Marketing, 25*, pp. 102–16.

Bazargan-Lari, M. (2004). Flexible versus fixed timetabling: A case study. *Journal of the Operational Research Society, 55*, pp. 123–31.

Bazerman, M.H., & Chugh, D. (2006). Decisions without blinders. *Harvard Business Review, 84*(1), pp. 88–97.

Bazzanella, C., & Damiano, R. (1999). The interactional handling of misunderstanding in everyday conversations. *Journal of Pragmatics, 31*, pp. 817–36.

Beaulieu, P.R. (1994). Commercial lenders' use of accounting information in interaction with source credibility. *Contemporary Accounting Research, 10*, pp. 557–85.

Bedny, G.Z., Karwowski, W., & Jeng, O.J. (2004). The situational reflection of reality in activity theory and the concept of situation awareness in cognitive psychology. *Theoretical Issues in Ergonomics Science, 5*, pp. 275–96.

Beilock, S.L., Wierenga, S.A., & Carr, T.H. (2002). Expertise, attention, and memory in sensorimotor skill execution: Impact of novel task constraints on dual-task performance and episodic memory. *Quarterly Journal of Experimental Psychology, 55*, pp. 1211–40.

Bell, D.E., Raiffa, H., & Tversky, A. (1988). *Decision Making: Descriptive, Normative, and Prescriptive Interactions*. Cambridge, MA: Harvard University Press.

Bellenkes, A.H., Wickens, C.D., & Kramer, A.F. (1997). Visual scanning and pilot expertise: The role of attentional flexibility and mental model development. *Aviation, Space, and Environmental Medicine, 68*, pp. 569–79.

Belsky, E.S., Case, K.E., & Smith, S.J. (2008). *Identifying, Managing and Mitigating Risks to Borrowers in Changing Mortgage and Consumer Credit Markets (UCC08-14)*. Boston, MA: Joint Center for Housing Studies.

Bereiter, C., & Scardamalia, M. (1989). Intentional learning as a goal of instruction. In L. Resnick (ed.), *Essays in Honour of Robert Glaser* (pp. 361–3). Hillsdale, NJ: Lawrence Erlbaum Associates.

Berger, J., Cohen, B.P., & Zelditch Jr, M. (1972). Status characteristics and social interaction. *American Sociological Review, 37*, pp. 241–55.

Berlyne, D.E. (1971). *Aesthetics and Psychobiology*. New York: Appleton-Century-Crofts.

Berto, R., Massaccesi, S., & Pasini, M. (2008). Do eye movements measured across high and low fascination photographs differ? Addressing Kaplan's fascination hypothesis. *Journal of Environmental Psychology, 28*, pp. 185–91.

Bertrams, R.I.V.F. (1979). The Tenerife aircrash litigation in the United States: Complex legal and practical problems and how they are solved. *Netherlands International Law Review, 26*, pp. 278–315.

Betsch, T., Haberstroh, S., Molter, B., & Glöckner, A. (2004). Oops, I did it again – Relapse errors in routinized decision making. *Organizational Behavior and Human Decision Processes, 93*, pp. 62–74.

Beutler, L.E., Hinton, R.M., Crago, M., & Collier, S. (1995). Evaluation of "fixed propensities" to commit sexual offences – A preliminary report. *Criminal Justice and Behavior, 22*, 284–94.

Biggs, S.F., Bedard, J.C., Gaber, B.G., & Linsmeier, T.J. (1985). The effects of task size and similarity on the decision behavior of bank loan officers. *Management Science, 31*, pp. 970–87.

Bills, A.G. (1935). Fatigue, oscillation, and blocks. *Journal of Experimental Psychology, 18*, pp. 562–73.

Bishop, C.M. (1995). *Neural Networks for Pattern Recognition.* Oxford: Oxford University Press.

Bitterman, M.E. (2006). Classical conditioning since Pavlov. *Review of General Psychology, 10*, pp. 365–76.

Blandford, A., and Wong, B.L.W. (2004). Situational awareness in emergency medical dispatch. *International Journal of Human-Computer Studies, 61*, pp. 421–52.

Blascovich, J., & Mendes, W.B. (2000). Challenge and threat appraisals: The role of affective cues. In J.P. Forgas (ed.), *Feeling and Thinking: The Role of Affect in Social Cognition* (pp. 59–82). New York: Cambridge University Press.

Blascovich, J., Seery, M.D., Mugridge, C.A., Norris, R.K., & Weisbuch, M. (2004). Predicting athletic performance from cardiovascular indexes of challenge and threat. *Journal of Experimental Social Psychology, 40*, pp. 683–8.

Bless, H., & Fiedler, K. (2006). Mood and the regulation of information processing and behavior. In J.P. Forgas (ed.), *Affect in Social Thinking and Behaviour* (pp. 65–84). New York: Psychology Press.

Blignaut, C.J.H. (1979). The perception of hazard: II. The contribution of signal detection to hazard perception. *Ergonomics, 22*, pp. 1177–83.

Blume, B.D., Ford, J.K., Baldwin, T.T., & Huang, J.L. (2010). Transfer of training: A meta-analytic review. *Journal of Management, 36*, pp. 1065–105.

Boeing Commercial Airplanes (2013). *2013 Pilot and Technical Outlook.* Seattle, WA: Author.

Bond, A.B., & Kamil, A.C. (2002). Visual predators select for crypticity and polymorphism in virtual prey. *Nature, 415*, pp. 609–13.

Bond, W.F., Schwartz, L.M., Weaver, K.R., Levick, D., Giuliano, M., & Graber, M.L. (2012). Differential diagnosis generators: An evaluation of currently available computer programs. *Journal of General Internal Medicine, 27*, pp. 213–19.

Boreham, N. (1995). Error analysis and expert – novice differences in medical diagnosis. In J.M. Hoc, P.C. Cacciabue, & E. Hollnagel (eds), *Expertise and Technology* (pp. 93–105). Hillsdale, NJ: Lawrence Erlbaum.

Boschen, A.C., & Jones, K.R. (2004). *Aviation Language Problem: Improving Pilot-controller Communication.* Paper presented at the International Professional Communication Conference, Minneapolis, Minnesota.

Boschker, M.S., & Barker, F.C. (2002). Inexperienced sport climbers might perceive and utilize new opportunities for action by merely observing a model. *Perceptual and Motor Skills, 95*, pp. 3–9.

Bourke, P., & Shanmugam, B. (1990). *An Introduction to Bank Lending.* Sydney: Addison-Wesley.

Bower, G.H. (1981). Mood and memory. *American Psychologist, 36*, pp. 129–48.

Bower, G.H. (1991). Mood congruity of social judgements. In J.P. Forgas (ed.), *Emotion and Social Judgements* (pp. 31–53). Oxford: Pergamon Press.

Bowling, S.R., Khasawneh, M.T., Kaewkuekool, S., Jiang, X., & Gramopadhye, A.K. (2008). Evaluating the effects of virtual training in an aircraft maintenance task. *International Journal of Aviation Psychology, 18*, pp. 104–16.

Bradlow, A., & Bent, T. (2008). Perceptual adaptation to non-native speech. *Cognition, 106*, pp. 707–29.

Brehmer, B. (1981). Models of diagnostic judgements. In J. Rasmussen & W.B. Rouse (eds), *Human Detection and Diagnosis of System Failures* (pp. 231–9). New York: Springer-Verlag.

Brennan, S.A., & Williams, M. (1995). The feeling of another's knowing: Prosody and filled pauses as cues to listeners about the metacognitive states of speakers. *Journal of Memory and Language, 34*, pp. 383–98.

Broadbent, K.D. (1957). Effects of noises of high and low frequency on behavior. *Ergonomics, 1*, pp. 21–9.

Broder, A., & Eichler, A. (2006). The use of recognition information and additional cues in inferences from memory. *Acta Psychologica, 121*, pp. 275–84.

Bronner, R. (1982). *Decision Making Under Time Pressure*. Lexington, MA: Lexington Books.

Browne, G.J., & Pitts, M.G. (2004). Stopping rule use during information search in design problems. *Organizational Behavior and Human Decision Processes, 95*, pp. 208–24.

Bruce, P.J. (2011). *Understanding Decision-Making Processes in Airline Operations Control*. Farnham: Ashgate.

Brunswik, E. (1952). *The Conceptual Framework of Psychology*. Chicago, IL: University of Chicago Press.

Brunswik, E. (1955). Representative design and probabilistic theory in a functional psychology. *Psychological Review, 62*, pp. 193–217.

Brunswik, E. (1956). *Perception and the Representative Design of Psychological Experiments*. Berkeley, CA: University of California Press.

Brunswik, E., Hammond, K.R., & Stewart, T.R. (2000). *The Essential Brunswik: Beginnings, Explications, and Applications*. New York: Oxford University Press.

Bucknall, C.E., Moran, F., Robertson, C., & Stevenson, R.D. (1988). Differences in hospital asthma management. *The Lancet, 331*, pp. 748–50.

Bureau d'Enquêtes et d'Analyses. (2012). *Final Report, Air France Flight AF447, Rio de Janeiro – Paris, June 1, 2009*. Paris: Author.

Burke, C.S., Stagl, K.C., Salas, E., Pierce, L. & Kendall, D. (2006). Understanding team adaption: A conceptual analysis and model. *Journal of Applied Psychology, 91*, pp. 1189–207.

Bushnell, P.J., Benignus, V.A., & Case, M.W. (2003). Signal detection behavior in humans and rats: A comparison with matched tasks. *Behavioural Processes, 64*, pp. 121–9.

Canter, D.V. (2000). Offender profiling and criminal differentiation. *Legal and Criminological Psychology, 5*, 23–46.

Canter, D.V., & Fritzon, K. (1998). Differentiating arsonists: A model of firesetting actions and characteristics. *Legal and Criminological Psychology, 3*, 73–96.

Canter, D.V., Coffey, T., Huntley, M., & Missen, C. (2000). Predicting serial killers' home base using a decision support system. *Journal of Quantitative Criminology, 16*, 457–78.

Carayon, P., & Gurses A.P. (2005). A human factors engineering conceptual framework of nursing workload and patient safety in intensive care units. *Intensive Critical Care Nursing, 21*, 284–301.

Carberry, S. (1990). *Plan Recognition in Natural Language Dialogue.* Cambridge, MA: MIT Press.

Caretta, T.R., Perry, D.C., & Ree, M.J. (1996). Prediction of situational awareness in F-15 pilots. *International Journal of Aviation Psychology, 6*, 21–41.

Caro, T.M. (1994). *Cheetahs of the Serengeti Plains: Group Living in an Asocial Species.* Chicago, IL: University of Chicago Press.

Carroll, J.S., & Rudolph, J.W. (2006). Design of high reliability organizations in health care. *Quality and Safety in Health Care, 15*(suppl 1), i4–i9.

Carruthers, B.G. (2012). What's haute in the sociology of finance? *Contemporary Sociology: A Journal of Reviews, 41*, 739–55.

Carter, M., & Beier, M.E. (2010). The effectiveness of error management training with working-aged adults. *Personnel Psychology, 63*, 641–75.

Carthey, J., de Leval, M.R., & Reason, J.T. (2001). The human factor in cardiac surgery: Errors and near misses in a high technology medical domain. *The Annals of Thoracic Surgery, 72*, 300–305.

Casey Jr, C.J. (1980). Variation in accounting information load: The effect on loan officers' predictions of bankruptcy. *Accounting Review, 55*, 36–49.

Casner, S.M., Geven, R.W., & Williams, K.T. (2013). The effectiveness of airline pilot training for abnormal events. *Human Factors, 55*, 477–85.

Cellier, J.M., Eyrolle, H., & Marine, C. (1997). Expertise in dynamic environments. *Ergonomics, 40*, 28–50.

Chan, A., & Payne, J. (2013). *Homicide in Australia: 2008–09 to 2009–10. National Homicide Monitoring Program Annual Report.* Canberra: Australian Institute of Criminology.

Chanthivong, A., Coleman, A.D.F., & Esho, N. (2003). *Report on Broker Originated Lending: Results of a Survey of Authorised Deposit-taking Institutions.* Canberra: Australian Prudential Regulation Authority (APRA).

Charness, N. (1979). Components of skill in bridge. *Canadian Journal of Psychology, 33*, 1–16.

Chase, W.G., & Simon, H.A. (1973a). *The Mind's Eye in Chess.* New York: Academic Press.

Chase, W.G., & Simon, H.A. (1973b). Perception in chess. *Cognitive Psychology, 1*, 55–81.

Chi, M.T., Glaser, R., & Farr, M.J. (1988). *The Nature of Expertise*. Hillsdale, NJ: Lawrence Erlbaum Associates.

Chi, M.T.H., Glaser, R., & Rees, E. (1982). Expertise in problem solving. In R. Sternberg (ed.), *Advances in the Psychology of Human Intelligence* (vol. 1, pp. 7–75). Hillsdale, NJ: Lawrence Erlbaum Associates.

Christenssen-Szalanski, J.J., & Bushyhead, J.R. (1981). Physician's use of probabilistic information in a real clinical setting. *Journal of Experimental Psychology: Human Perception and Performance, 7*, pp. 928–36.

Chu, P.C., & Spires, E.E. (2001). Does time constraint on users negate the efficacy of decision support systems? *Organizational Behavior and Human Decision Processes, 85*, pp. 226–49.

Cialdini, R.B., Kallgren, C.A., & Reno, R.R. (1991). A focus theory of normative conduct: A theoretical refinement and reevaluation of the role of norms in human behavior. *Advances in Experimental Social Psychology, 24*, pp. 1–243.

Clancey, W.J. (1997). *Situated Cognition: On Human Knowledge and Computer Representations*. Cambridge: Cambridge University Press.

Clark, H.H. (1996). *Using language*. Cambridge: Cambridge University Press.

Clark, H.H., & Brennan, S.E. (1991). Grounding in communication. In H.H. Clark & S.E. Brennan (eds), *Perspectives on Socially Shared Cognition* (pp. 127–49). Washington, DC: American Psychological Association.

Clark, H.H., & Krych, M.A. (2004). Speaking while monitoring addressees for understanding. *Journal of Memory and Language, 50*, pp. 62–81.

Clark, H.H., & Schaefer, E.F. (1987). Collaborating on contributions to conversations. *Language and Cognitive Processes, 2*, pp. 19–41.

Clark, H.H., & Wilkes-Gibbs, D. (1986). Referring as a collaboration process. *Cognition, 22*, pp. 1–39.

Clarke, C.M., & Garrett, M.F. (2004). Rapid adaptation to foreign-accented English. *Journal of the Acoustical Society of America, 116*, pp. 3647–58.

Clausen, J., Larsen, A., Larsen, J., & Rezanova, N.J. (2010). Disruption management in the airline industry – Concepts, models and methods. *Computers & Operations Research, 37*, pp. 809–21.

Clore, G.L., & Parrott, G. (1991). Moods and their vicissitudes: Thoughts and feelings as information. In J.P. Forgas (ed.), *Emotion and Social Judgements* (pp. 107–23). Oxford: Pergamon.

Coderre, S., Mandin, H., Harasym, P.H., & Fick, G.H. (2003). Diagnostic reasoning strategies and diagnostic success. *Medical Education, 37*, pp. 695–703.

Cohen, M.S., & Freeman, J.T. (1996). Thinking naturally about uncertainty. *Proceedings of the Human Factors and Ergonomics Society 40th Annual Meeting* (pp. 179–83). Santa Monica, CA: HFES.

Cohen, M.S., Freeman, J.T., & Wolf, S. (1996). Metarecognition in time-stressed decision making: Recognizing, critiquing, and correcting. *Human Factors, 38*, pp. 206–19.

Cohen, P.R., & Levesque, H.J. (1990). Rational interactions the basis for communication. In P.R. Cohen, J. Morgan, & M.E. Pollack (eds), *Intentions in Communication* (pp. 221–56). Cambridge, MA: MIT Press.

Cohen, P.R., Morgan, J., & Pollack, M.E. (1990). *Intentions in Communication*. Cambridge, MA: MIT Press.

Cole, R.A. (1998). The importance of relationships to the availability of credit. *Journal of Banking and Finance, 22*, pp. 959–77.

Colton, R.C. (1985). *The Civil War in the Western Territories*. Norman, OK: University of Oklahoma Press.

Connell, R.W. (1987). *Gender and Power*. Stanford, CA: Stanford University Press.

Connell, R.W. (1994). Psychoanalysis on masculinity. In H. Brod & M. Kaufman (eds), *Theorizing Masculinities* (pp. 11–38). Thousand Oaks, CA: Sage Publications.

Cooper, G., Tindall-Ford, S., Chandler, P., & Sweller, J. (2001). Learning by imagining. *Journal of Experimental Psychology: Applied, 7*, p. 68.

Cosby, K.S., & Croskerry, P. (2004). Profiles in patient safety: Authority gradients in medical error. *Academic Emergency Medicine, 11*, pp. 1341–5.

Craig, C., Klein, M.I., Griswold, J., Gaitonde, K., McGill, T., & Halldorsson, A. (2012). Using cognitive task analysis to identify critical decisions in the laparoscopic environment. *Human Factors, 54*, pp. 1025–39.

Crofts, J.F., Fox, R., Ellis, D., Winter, C., Hinshaw, K., & Draycott, T.J. (2008). Observations from 450 shoulder dystocia simulations: Lessons for skills training. *Obstetrics & Gynecology, 112*, pp. 906–12.

Croskerry, P. (2002). Achieving quality in clinical decision making: Cognitive strategies and detection of bias. *Academic Emergency Medicine, 9*, pp. 1184–204.

Croskerry, P. (2009a). A universal model of diagnostic reasoning. *Academic Medicine, 84*, pp. 1022–8.

Croskerry, P. (2009b). Clinical cognition and diagnostic error: Applications of a dual process model of reasoning. *Advances in Health Sciences Education, 14*, pp. 27–35.

Curci, A., & Luminet, O. (2009). Flashbulb memories for expected events: A test of the emotional-integrative model. *Applied Cognitive Psychology, 23*, pp. 98–114.

Curnin, S.W. and Owen, C. (2013). Obtaining information in emergency management: A case study from an Australian emergency operations centre. *International Journal of Human Factors and Ergonomics, 2*, pp. 131–58.

Cushing, S. (1997). *Fatal Words: Communication Clashes and Aircraft Crashes*. Chicago, IL: University of Chicago Press.

Cutting, A.L., & Dunn, J. (2002). The cost of understanding other people: Social cognition predicts young children's sensitivity to criticism. *Journal of Child Psychology and Psychiatry, 43*, pp. 849–60.

Czeisler, C.A. (2009). Medical and genetic differences in the adverse impact of sleep loss on performance: Ethical considerations for the medical profession. *Transactions of the American Clinical and Climatological Association, 120*, p. 249.

Daglish, T. (2009). What motivates a subprime borrower to default? *Journal of Banking and Finance, 33*, pp. 681–93.

Damasio, A.R. (1994). *Descartes' Error*. New York: Grosset/Putnam.

Davies, D.R. and Parasuraman, R. (1982). *The Psychology of Vigilance*. London: Academic Press.

Davies, D.R., Matthews, G., Stammers, R.B., & Westerman, S.J. (2013). *Human Performance: Cognition, Stress and Individual Differences*. East Sussex: Psychology Press.

Dawes, R.M. (1979). The robust beauty of improper linear models in decision making. *American Psychologist, 34*, pp. 571–82.

Dawes, R.M., & Corrigan, B. (1974). Linear models in decision making. *Psychological Bulletin, 81*, pp. 95–106.

de Fockert, J.W., Rees, G., Frith, C.D., & Lavie, N. (2001). The role of working memory in visual selective attention. *Science, 291*, pp. 1803–6.

de Groot, A.D. (1965). *Thought and Choice in Chess*. The Hague: Moutain.

de Groot, A.D. (1966). Perception and memory versus thought: Some old ideas and recent findings. In B. Kleinmuntz (ed.), *Problem solving* (pp. 19–50). New York: John Wiley.

de Leval, M.R., Carthey, J., Wright, D.J., Farewell, V.T., & Reason, J.T. (2000). Human factors and cardiac surgery: A multicultural study. *The Journal of Thoracic and Cardiovascular Surgery, 119*, pp. 661–72.

Deng, A. (2013). The thin blue line getter thinner, *China Daily*. Available at: https://www.chinadailyasia.com/news/2013-04-26/content_15073398.html.

Denzin, N., & Lincoln, Y. (eds.). (2004). *Handbook of Qualitative Research*. Thousand Oaks, CA: Sage.

Deutsch, M., & Gerard, H. B. (1955). A study of normative and informational social influences upon individual judgment. *The Journal of Abnormal and Social Psychology, 51*, 629.

Didierjean, A., & Fernand, G. (2008). Sherlock Holmes – A expert's view of expertise. *British Journal of Psychology, 99*, pp. 109–25.

Dolan, P., Jones-Lee, M., & Loomes, G. (1995). Risk-risk versus standard gamble procedures for measuring health state utilities. *Applied Economics, 27*, pp. 1103–11.

Donchin, Y., & Seagull, F.J. (2002). The hostile environment of the intensive care unit. *Current Opinion in Critical Care, 8*, pp. 316–320.

Donchin, Y., Gopher, D., Olin, M., Badihi, Y., Biesky, M.R., Sprung, C.L., ... & Cotev, S. (1995). A look into the nature and causes of human errors in the intensive care unit. *Critical Care Medicine, 23*, pp. 294–300.

Douglas, J.E., Burgess, A.W., Burgess, A.G., & Resller, R.K. (2013). *Crime Classification Manual* (3rd ed.). New York: Wiley.

Drach-Zahavy, A., & Somech, A. (2001). Understanding team innovation: The role of team processes and structures. *Group Dynamics: Theory, Research, and Practice, 5*, p. 111.

Drach-Zahavy, A., & Somech, A. (2011). Narrowing the gap between safety policy and practice: The role of nurses' implicit theories and heuristics. In E. Rowley & J. Waring (eds), *A Socio-Cultural Perspective on Patient Safety* (pp. 51–70). Farnham: Ashgate Publishing.

Drohan, B. (2006). Carl von Clausewitz, his trinity, and the War of 1812, Part Two. *Journal of Slavic Military Studies, 19*, 515–42.

Dror, I.E., Stevenage, S.V., & Ashworth, A.R.S. (2008). Helping the cognitive system learn: Exaggerating distinctiveness and uniqueness. *Applied Cognitive Psychology, 22*, pp. 573–84.

Drummond, K., & Hopper, R. (1991). Misunderstanding and its remedies: Telephone miscommunication. In N. Coupland, H. Giles, & J.M. Wiemann (eds), *"Miscommunication" and Problematic Talk* (pp. 301–14). Newbury Park, CA: Sage.

Drury, C.G., & Ma, J. (2002). *Language Error Analysis: Report on Literature of Aviation Language, Errors and Analysis of Error Databases*. Washington, DC: Federal Aviation Administration (FAA).

Dukas, R., & Kamil, A.C. (2000). The cost of limited attention in blue jays. *Behavioral Ecology, 11*, pp. 502–6.

Dukas, R., & Kamil, A.C. (2001). Limited attention: The constraint underlying search image. *Behavioral Ecology, 12*, pp. 192–9.

Durso, F.T., & Dattel, A.R. (2004). SPAM: The real-time assessment of SA. In S. Banbury & S. Tremblay (eds), *A Cognitive Approach to Situational Awareness: Theory and Application* (pp. 137–54). Aldershot: Ashgate.

Durso, F.T., and Sethumadhavan, A. (2008). Situational awareness: Understanding dynamic environments. *Human Factors, 50*, pp. 442–8.

Easterbrook, J.A. (1959). The effect of emotion on cue utilization and the organization of behavior. *Psychological Review, 66*, pp. 183–201.

Ebbesen, E.B., & Konecni, V.J. (1975). Decision making and information integration in the courts: The setting of bail. *Journal of Personality and Social Psychology, 32*, pp. 805–21.

Ebenbach, D.H., & Moore, C.F. (2000). Incomplete information, inferences, and individual differences: The case of environmental judgments. *Organizational Behavior and Human Decision Processes, 81*, pp. 1–27.

Eddy, D. (1982). Probabilistic reasoning in clinical medicine: Problems and opportunities. In D. Kahneman, P. Slovic., & A. Tversky (eds), *Judgement Under Uncertainty: Heuristics and Biases* (pp. 249–67). Cambridge: Cambridge University Press.

Eddy, D. (1988). Variations in physician practice: The role of uncertainty. In J. Dowie & A.S. Elstein (eds), *Professional Judgement: A Reader in Clinical Decision Making* (pp. 200–211). Cambridge: Cambridge University Press.

Edmondson, A.C. (2003). Speaking up in the operating room: How team leaders promote learning in interdisciplinary action teams. *Journal of Management Studies, 40*, pp. 1419–52.

Einbinder, J.S., Klein, D.A., & Safran, C.S. (1997). Making effective referrals: A knowledge-management approach. In *Proceedings of the AMIA Annual Fall Symposium* (p. 330). Washington, DC: American Medical Informatics Association.

Einhorn, H. (1974). Expert judgement: Some necessary conditions and an example. *Journal of Applied Psychology, 59*, pp. 562–71.

Elsbach, K. D., & Barr, P. S. (1999). The effects of mood on individuals' use of structured decision protocols. *Organization Science, 10*, pp. 181–98.

Elsesser, K., Heuschen, I., Pundt, I., & Sartory, G. (2006). Attentional bias and evoked heart-rate response in specific phobia. *Cognition and Emotion, 20*, pp. 1092–107.

Ely, J.W., Graber, M.L., & Croskerry, P. (2011). Checklists to reduce diagnostic errors. *Academic Medicine, 86*, pp. 307–13.

Endsley, M.R. (1995a). Toward a theory of situational awareness in dynamic systems. *Human Factors, 37*, pp. 32–64.

Endsley, M.R. (1995b). Measurement of situational awareness in dynamic systems. *Human Factors, 37*, pp. 65–84.

Endsley, M.R. (1996). Automation and situational awareness. In R. Parasuraman & M. Mouloua (eds), *Automation and Human Performance: Theory and Application* (pp. 163–81). Mahwah, NJ: Lawrence Erlbaum.

Endsley, M.R., & Bolstad, C.A. (1994). Individual differences in pilot situational awareness. *The International Journal of Aviation Psychology, 4*, pp. 241–64.

Endsley, M.R., & Kiris, E.O. (1995). The out-of-the-loop performance problem and level of control in automation. *Human Factors, 37*, pp. 381–94.

Engeström, Y. (2004). The new generation of expertise. In H. Rainbird, A. Fuller, & A. Munro (eds), *Workplace Learning in Context* (pp. 145–66). New York: Routledge.

Enis, C.R. (1995). Expert-novice judgements and new cue sets: Process versus outcome. *Journal of Economic Psychology, 16*, pp. 641–2.

Ericsson, K.A. (2005). Superior decision making as an integral quality of expert performance: Insights into the mediating mechanisms and their acquisition through deliberate practice. In D.A. Montgomery, R. Lipshitz, & B. Brehmer (eds), *How Professionals Make Decisions* (pp. 135–61). Mahwah, NJ: Lawrence Erlbaum Associates.

Ericsson, K.A., & Kintsch, W. (1995). Long-term working memory. *Psychological Review, 102*, pp. 211–45.

Ericsson, K.A., & Lehmann, A.C. (1996). Expert and exceptional performance: Evidence of maximal adaptation to task constraints. *Annual Review of Psychology, 47*, pp. 273–305.

Ericsson, K.A., Krampe, R., & Tesch-Römer, C. (1993). The role of deliberate practice in the acquisition of expert performance. *Psychological Review, 100*, pp. 363–406.

Euromonitor International. (2009). *Consumer Lending in Australia.* Sydney: Author.

Evans, J.S. (2008). Dual-processing accounts of reasoning, judgement, and social cognition. *Annual Review of Psychology, 59*, pp. 255–78.

Fabrikant, S. I. (2005, March). Towards an understanding of geovisualization with dynamic displays: Issues and prospects. In *AAAI Spring Symposium: Reasoning with Mental and External Diagrams: Computational Modeling and Spatial Assistance* (pp. 6–11). Stanford, CA: Stanford University.

Fegyveresi, A.E. (1997). Vocal cues and pilot/ATC communications. In R.S. Jensen (ed.), *Proceedings of the Ninth International Symposium on Aviation Psychology* (pp. 81–4). Columbus, OH: Aviation Psychology Laboratory, Ohio State University.

Festa, M., de Pont, J., Crone, L., Schell, D., & Wiggins, M. (2009). Use of forensic (cognitive) interview technique to identify important handover information in post-operative paediatric cardiac surgery patients. In *Proceedings of 5th World Congress of Paediatric Cardiology and* Cardiac *Surgery*, Australia.

Festinger, L. (1962). *A Theory of Cognitive Dissonance* (vol. 2). Stanford, CA: Stanford University Press.

Fine, G.A. (1996). Justifying work: Occupational rhetorics as resources in restaurant kitchens. *Administrative Science Quarterly, 41*, pp. 90–115.

Fineman, S., & Sturdy, A. (1999). The emotions of control: A qualitative exploration of environmental regulation. *Human Relations, 52*, pp. 631–63.

Finkbeiner, M., & Forster, K. I. (2007). Attention, intention, and domain-specific processing. *Trends in Cognitive Sciences, 12*, pp. 59–64.

Fisher, A., & Fonteyn, M. (1995). A exploration of an innovative methodological approach for examining nurses' heuristic use in clinical practice. *Scholarly Inquiry for Nursing Practice: An International Journal, 9*, pp. 263–76.

Fisher, D., & Pollatsek, A. (2007). Novice driver crashes: Failure to divide attention or failure to recognize risks. In A.F. Kramer, D.A. Wiegmann, & A. Kirlik (eds), *Attention: From Theory to Practice* (pp. 134–56). Oxford: Oxford University Press.

Fitts, P.M. (1966). Cognitive aspects of information processing: III. Set for speed versus accuracy. *Journal of Experimental Psychology, 71*, pp. 849–57.

Flach, J.M. (1995). Situational awareness: Proceed with caution. *Human Factors, 37*, pp. 149–57.

Flege, J.E. (1995). Second language speech learning: Theory, findings and problems. In W. Strange (ed.), *Speech Perception and Linguistic Experience: Issues in Cross-language Research* (pp. 233–77). Timonium, MD: York Press.

Flege, J.E., Munro, M.J., & MacKay, I.R.A. (1995). Effects of age of second-language learning on the production of English consonants. *Speech Communication, 16*, pp. 1–26.

Flin, R., O'Connor, P., & Crichton, M. (2008). *Safety at the Sharp End: A Guide to Non-Technical Skills*. Aldershot: Ashgate.

Foley, M., & Hart, A. (1990). Expert-novice differences and knowledge elicitation. In R.R. Hoffman (ed.), *The Psychology of Expertise* (pp. 233–44). New York: Springer-Verlag.

Forgas, J.P. (ed.). (2012). *Affect in Social Thinking and Behavior*. New York: Psychology Press.

Forgas, J.P. (1995). Mood and judgement: The affect infusion model (AIM). *Psychological Bulletin, 117*, pp. 1–28.

Forgas, J.P. (2001). The affect infusion model (AIM): An integrative theory of mood effects on cognition and judgement. In L.L. Martin & G.L. Clore (eds), *Theories of Mood and Cognition: A User's Guidebook* (pp. 99–134). Mahwah, NJ: Lawrence Erlbaum.

Forgas, J.P., & East, R. (2008). On being happy and gullible: Mood effects on skepticism and the detection of deception. *Journal of Experimental Social Psychology, 44*, pp. 1362–7.

Forgas, J.P., Laham, S., & Vargas, P. (2005). Mood effects on eyewitness memory: Affective influences on susceptibility to misinformation. *Journal of Experimental Social Psychology, 41*, pp. 574–88.

Friedrich, J. (1993). Primary error detection and minimisation (PEDMIN) in social cognition: A reinterpretation of the confirmation bias phenomenon. *Psychological Review, 100*, pp. 298–319.

Gaba, D.M., Howard, S.K., & Small, S.D. (1995). Situational awareness in anesthesiology. *Human Factors, 37*, pp. 20–31.

Gadd, C.S. (1995). A theory of the multiple roles of diagnosis in collaborative problem solving discourse. In *Proceedings of the Seventeenth Annual Conference of the Cognitive Science Society* (pp. 352–57). Hillsdale, NJ: Erlbaum.

Galinsky, T.L., Rosa, R.R., Warm, J.S., & Dember, W.N. (1993). Psychophysical determinants of stress in sustained attention. *Human Factors, 35*, pp. 603–14.

Garcia-Retamero, R., Hoffrage, U., & Dieckmann, A. (2007). When one cue is not enough: Combining fast and frugal heuristics with compound cue processing. *Quarterly Journal of Experimental Psychology, 60*, pp. 1197–215.

Gattie, G.J., & Bisantz, A.M. (2006). The effects of integrated cognitive feedback components and task conditions on training in a dental diagnosis task. *International Journal of Cognitive Ergonomics, 36*, pp. 485–97.

Gegenfurtner, A., Lehtinen, E., & Säljö, R. (2011). Expertise differences in the comprehension of visualizations: A meta-analysis of eye-tracking research in professional domains. *Educational Psychology Review, 23*, pp. 523–52.

Gibson, C. (1983, Summer). Financial ratios as perceived by commercial loan officers. *Akron Business and Economic Review*, pp. 23–7.

Gibson, W.H., Megaw, E.D., Young, M.S., & Lowe, E. (2006). A taxonomy of human communication errors and application to railway track maintenance. *Cognition, Technology & Work, 8*, pp. 57–66.

Gigerenzer, G. (2002). *Calculated Risks*. New York: Simon & Schuster.

Gigerenzer, G., & Brighton, H. (2009). Homo heuristicus: Why biased minds make better inferences. *Topics in Cognitive Science, 1*, pp. 107–43.

Gigerenzer, G., & Goldstein, D.G. (1996). Reasoning the fast and frugal way: Models of bounded rationality. *Psychological Review, 103*, pp. 650–69.

Gigerenzer, G., & Goldstein, D.G. (1999). Betting on one good reason: The take-the-best heuristic. In G. Gigerenzer, P.M. Todd and the ABC Research Group, *Simple Heuristics that Make Us Smart* (pp. 75–95). New York: Oxford University Press.

Gigerenzer, G., Todd, P.M., & the ABC Research Group (1999). *Simple Heuristics that Make Us Smart*. New York: Oxford University Press.

Gil-Alana, L.A., Barros, C.P., & de Araujo, A.F.J. (2012). Aircraft accidents in Brazil. *International Journal of Sustainable Transportation, 6*, pp. 111–26.

Gildea, K.M., Schneider, T R., & Shebilske, W.L. (2007). Stress appraisals and training performance on a complex laboratory task. *Human Factors, 49*, pp. 745–58.

Gill, A.S., Flaschner, A.B., & Shachar, M. (2006). Factors that affect the trust of business clients in their banks. *International Journal of Bank Marketing, 24*, pp. 384–405.

Gill, C.J., Sabin, L., & Schmid, C.H. (2005). Why clinicians are natural Bayesians. *British Medical Journal, 330*, pp. 1080–83.

Gjerberg, E. (2003). Women doctors in Norway: The challenging balance between career and family life. *Social Science and Medicine, 57*, pp. 1327–41.

Glaser, R., Chi, M.T., & Farr, M.J. (eds). (1988). *The Nature of Expertise* (pp. xv–xxviii). Hillsdale, NJ: Lawrence Erlbaum Associates.

Goel, N., Rao, H., Durmer, J.S., & Dinges, D.F. (2009, September). Neurocognitive consequences of sleep deprivation. In *Seminars in Neurology* (Vol. 29, No. 4, p. 320). Washington, DC: National Institute of Health Public Access.

Goffman, E., (1960). *The Presentation of Self in Everyday Life*. New York: Doubleday.

Goffman, E. (1974). *Frame Analysis: An Essay on the Organization of Experience*. Boston, MA: Harvard University Press.

Gohar, A., Adams, A., Gertner, E., Sackett-Lundeen, L., Heitz, R., Engle, R., Haus, E. & Bijwadia, J. (2009). Working memory capacity is decreased in sleep-deprived internal medicine residents. *Journal of Clinical Sleep Medicine, 5*, p. 191.

Goldstein, D.G., & Gigerenzer, G. (2002). Models of ecological rationality: The recognition heuristic. *Psychological Review, 109*, pp. 75–90.

Gonzalez, C. (2004). Learning to make decisions in dynamic environments: Effects of time constraints and cognitive abilities. *Human Factors, 46*, pp. 449–60.

Gonzalez, C., & Brunstein, A. (2009). Training for emergencies. *Journal of Trauma, Injury, Infection & Critical Care, 67*(2 Suppl), S100–S105.

Gonzalez, C., & Thomas, R.P. (2008). Effects of automatic detection on dynamic decision making. *Journal of Cognitive Engineering and Decision Making*, *2*, pp. 328–48.

Goode, J.H. (2003). Are pilots at risk of accidents due to fatigue? *Journal of Safety Research*, *34*, pp. 309–13.

Graber, M.L., Franklin, N., & Gordon, R. (2005). Diagnostic error in internal medicine. *Archives of Internal Medicine*, *165*(13), pp. 1493–9.

Gray, W.D. (2000). The nature and processing of errors in interactive behavior. *Cognitive Science*, *24*, pp. 205–48.

Green, J., Booth, C.E., & Biderman, M.D. (2001). Cluster analysis of burglary M/Os. In M. Godwin (ed.), *Criminal Psychology and Forensic Technology: A Collaborative Approach to Effective Profiling* (pp. 153–66). Boca Raton, FL: CRC Press.

Greene, M.R., & Oliva, A. (2009). The briefest of glances: The time course of natural scene understanding. *Psychological Science*, *20*, pp. 464–72.

Grice, H.P. (1989). *Studies in the Way of Words*. Cambridge, MA: Harvard University Press.

Grove, W.M., & Meehl, P.E. (1996). Comparative efficiency of informal (subjective, impressionistic) and formal (mechanical, algorithmic) prediction procedures. *Psychology, Public Policy, and Law*, *2*, pp. 293–323.

Groves, M., O'Rourke, P., & Alexander, H. (2003). The clinical reasoning characteristics of diagnostic experts. *Medical Teacher*, *25*, pp. 308–13.

Grubin, D., Kelly, P., & Ayis, A. (1997). *Linking Serious Sexual Assaults*. London: Home Office Police Department, Police Research Group.

Guba, E.G. (1981). Criteria for assessing the trustworthiness of naturalistic inquiries. *Educational Communication and Technology Journal*, *29*, pp. 75–92.

Hahn, M., Lawson, R., & Lee, Y.G. (1992). The effects of time pressure and information load on decision quality. *Psychology and Marketing*, *9*, pp. 365–78.

Hales, B., Terblanche, M., Fowler, R., & Sibbald, W. (2008). Development of medical checklists for improved quality of patient care. *International Journal for Quality in Health Care*, *20*, pp. 22–30.

Hamilton, P., Hockey, G.R.J., & Rejman, M. (1977). The place of the concept of activation in human information processing theory. In S. Dormic (ed.), *Attention and Performance* (pp. 463–86). New York: Academic Press.

Hammond, K.R., Frederick, E.N., Robillard, N., & Victor, D. (1989). Application of cognitive theory to the student-teach dialogue. In D.A. Evans & V.L. Patel (eds), *Cognitive Science in Medicine: Biomedical Modelling* (pp. 173–210). Cambridge, MA: The MIT Press.

Hamra, J., Hossain, L., Owen, C. & Abbasi, A. (2012). Effects of networks on learning during emergency events. *Disaster Prevention and Management*, *21*, pp. 584–98.

Hanna, J.E., Tanenhaus, M.K., & Trueswell, J.C. (2003). The effects of common ground and perspective on domains of referential interpretation. *Journal of Memory and Language*, *39*, pp. 1–20.

Harhoff, D., & Körting, T. (1998). Lending relationships in Germany–Empirical evidence from survey data. *Journal of Banking and Finance, 22*, pp. 1317–53.

Hastie, R., & Dawes, R.M. (2001). *Rational Choice in an Uncertain World*. London: Sage.

Hawke, A., & Heffernan, T. (2006). Interpersonal liking in lender-customer relationships in the Australian banking sector. *International Journal of Bank Marketing, 24*, pp. 140–57.

Hawn, C. (2009). Take two aspirin and tweet me in the morning: How Twitter, Facebook, and other social media are reshaping health care. *Health Affairs, 28*(2), pp. 361–8.

Hayes, B.K., & Heit, E. (2004). Why learning and development can lead to poorer recognition memory. *Trends in Cognitive Science, 8*, 337–9.

Hazelwood, R.R., & Napier, M.R. (2004). Crime scene staging and its detection. *Criminology & Penology, 48*, pp. 744–59.

Head, H. (1920). *Studies in Neurology*. Oxford: Oxford University Press.

Head, H. (1923). The conception of nervous and mental energy II. Vigilance: A physiological state of the nervous system. *British Journal of Psychology, 14*, pp. 126–47.

Head, J., & Helton, W.S. (2012). Natural scene stimuli and lapses of sustained attention. *Consciousness and Cognition, 21*, pp. 1617–25.

Helmreich, R.L. (1994). Anatomy of a system accident: The crash of Avianca Flight 052. *International Journal of Aviation Psychology, 4*, pp. 265–84.

Helmreich, R.L. & Foushee, H.C. (1993). Why crew resource management? Empirical and theoretical bases of human factors training in aviation. In E.L. Weiner, B.G. Kanki, & R.L. Helmreich (eds), *Cockpit Resource Management* (pp. 3–46). San Diego, CA: Academic Press.

Helsen, W.F., & Starkes, J.L. (1999). A new training approach to complex decision making for police officers in potentially dangerous interventions. *Journal of Criminal Justice, 27*(5), pp. 395–410.

Helton, W.S. (2008). Expertise acquisition as sustained learning in humans and other animals: Commonalities across species. *Animal Cognition, 11*, pp. 99–107.

Helton, W.S., & Russell, P.N. (2011). Working memory load and the vigilance decrement. *Experimental Brain Research, 212*, pp. 429–37.

Helton, W.S., & Russell, P.N. (2012). Brief mental breaks and content-free cues may not keep you focused. *Experimental Brain Research, 219*, 37–46.

Helton, W.S., & Russell, P.N. (2013). Visuospatial and verbal working memory load: Effects on visuospatial vigilance. *Experimental Brain Research, 224*, pp. 429–36.

Helton, W.S., & Warm, J.S. (2008). Signal salience and the mindlessness theory of vigilance. *Acta Psychologica, 129*, pp. 18–25.

Helton, W.S., Dember, W.N., Warm, J.S., & Matthews, G. (1999). Optimism, pessimism, and false failure feedback: Effects on vigilance performance. *Current Psychology, 18*, pp. 311–25.

Helton, W.S., Weil, L., Middlemiss, A., & Sawers, A. (2010). Global interference and spatial uncertainty in the Sustained Attention to Response Task (SART). *Consciousness and Cognition, 19*, pp. 77–85.

Helton, W.S., Shaw, T., Warm, J.S., Matthews, G., & Hancock, P.A. (2008). Effects of warned and unwarned demand transitions on vigilance performance and stress. *Anxiety, Stress and Coping, 21*, pp. 173–84.

Helton, W.S., Hollander, T.D., Warm, J.S., Matthews, G., Dember, W.N., Wallaart, M., Beauchamp, G., Parasuraman, R., & Hancock, P.A. (2005). Signal regularity and the mindlessness model of vigilance. *British Journal of Psychology, 96*, pp. 249–61.

Henley, I., & Daly, W. (2004). Teaching non-native English speakers: Challenges and strategies. In M.A. Turney (ed.), *Tapping Diverse Talent in Aviation: Culture, Gender and Diversity* (pp. 21–43). Aldershot: Ashgate.

Herbig, B., Büssing, A., & Ewert, T. (2001). The role of tacit knowledge in the work context of nursing. *Journal of Advanced Nursing, 34*(5), pp. 687–95.

Hershey, D.A., Walsh, D.A., Read, S.J., & Chulef, A.S. (1990). The effects of expertise on financial problem solving: Evidence for goal-directed, problem solving scripts. *Organizational Behavior and Decision Processes, 46*, pp. 77–101.

Hines, S., Luna, K., Lofthus, J., Marquardt, M., & Stelmokas, D. (2008). *Becoming a High Reliability Organization: Operational Advice for Hospital Leaders.* Rockville, MD: Agency for Healthcare Research and Quality.

Hirst, G., McRoy, S., Heeman, P., Edmonds, P., & Horton, D. (1994). Repairing conversational misunderstandings and non-understandings. *Speech Communication, 15*, pp. 213–29.

Hirst, W., & Kalmar, D. (1987). Characterizing attentional resources. *Journal of Experimental Psychology: General, 116*, pp. 68–81.

Hitt, M.A., Beamish, P.W., Jackson, S.E., & Mathieu, J.E. (2007). Building theoretical and empirical bridges across levels: Multilevel research in management. *Academy of Management Journal, 50*, pp. 1385–99.

Hoc, J.M., Amalberti, R., & Boreham, N. (1995). Human operator expertise in diagnosis, decision-making and time management. In E. Hollnagel, P.C. Cacciabue & J.M. Hoc (eds), *Expertise and Technology: Cognition and Human-computer Cooperation* (pp. 19–42). Hillsdale, NJ: Lawrence Erlbaum Associates, Inc.

Hockey, G.R. (1970). Effect of loud noise on attentional selectivity. *The Quarterly Journal of Experimental Psychology, 22*(1), pp. 28–36.

Hoffman, R.R., Shadbolt, N.R., Burton, A.M., & Klein, G. (1995). Eliciting knowledge from experts: A methodological analysis. *Organizational Behavior and Human Decision Processes, 62*, pp. 129–58.

Hogarth, R.M. (1987). *Judgement and Choice* (2nd ed.). Chichester: John Wiley and Sons.

Holland, D., Lachicotte, D., Skinner, D., & Cain, C. (1998). *Identity and Agency in Cultural Worlds.* Cambridge, MA: Harvard University Press.

Hollnagel, E., & Woods, D.D. (2005). *Joint Cognitive Systems: Foundations of Cognitive Systems Engineering*. Boca Raton, FL: Taylor & Francis.

Howell, W.C., & Fleishman, E. (1982). *Information Processing and Decision Making*. Hillsdale, NJ: Lawrence Erlbaum.

Innes, J., & Lyon, R.A. (1994). A simulated lending decision with external management audit reports. *Accounting, Auditing and Accountability Journal*, *7*, pp. 73–93.

International Civil Aviation Organization (ICAO). (2009). *Guidelines for Aviation English Training Programs*. Montreal: Author

Itokawa, H. (2000). The mental state of an airline pilot as a machine operator. *International Association of Traffic and Safety Sciences Review*, *26*, pp. 48–56.

Ivancic, K., & Hesketh, B. (2000). Learning from errors in a driving simulation: Effects on driving skill and self-confidence. *Ergonomics*, *43*, pp. 1966–84.

Jackson, R.C., Warren, S., & Abernethy, B. (2006). Anitipcation skill and susceptibility to deceptive movement. *Acta Psychologica*, *123*, pp. 355–71.

Jackson, R.R., Pollard, S.D., & Cerveira, A.M. (2002). Opportunistic use of cognitive smokescreens by araneophagic jumping spiders. *Animal Cognition*, *5*, pp. 147–57.

Jancowicz, A.D., & Hisrich, R.D. (1987). Intuition in small business lending. *Journal of Small Business Management*, *2*, 45–52.

Jensen, R.S., Chubb, G.P., Adrion-Kochan, J., Kirkbride, L.A., & Fisher, J. (1995). Aeronautical decision making in general aviation: New intervention strategies. In R. Fuller, N. Johnston, & N. McDonald (eds), *Human Factors in Aviation Operations* (pp. 5–10). Aldershot: Ashgate.

Jentsch, F., Bowers, C., & Salas, E. (2001). What determines whether observers recognise targeted behaviours in modeling displays? *Human Factors*, *43*, pp. 496–507.

John, Y.J., Bullock, D., Zikopoulos, B., & Barbas, H. (2013). Anatomy and computational modeling of networks underlying cognitive-emotional interaction. *Frontiers in Human Neuroscience*, *7*(101), pp. 1–26.

Joint Commission on Accreditation of Healthcare Organizations. (2011). *Sound the Alarm: Managing Physiological Monitoring Systems*. Oakbrook Terrace, IL: Author.

Jones, D.J., & Endsley, M.R. (2000). Overcoming representational errors in complex environments. *Human Factors*, *42*, pp. 367–78.

Jones, R.K. (2003). Miscommunication between pilots and air traffic control. *Language Problems & Language Planning*, *27*, pp. 233–48.

Jordan, T.R., & Sergeant, P. (2000). Effects of distance on visual and audio-visual speech recognition. *Language and Speech*, *43*, pp. 107–24.

Julisch, K. (2003). Clustering intrusion detection alarms to support root cause analysis. *ACM Transactions on Information and System Security (TISSEC)*, *6*, pp. 443–71.

Kaber, D.B., & Endsley, M.R. (1997). Out-of-the-loop performance problems and the use of intermediate levels of automation for improved control system functioning and safety. *Process Safety Progress, 16*, pp. 126–31.

Kahneman, D. (1973). *Attention and Effort*. Englewood Cliffs, NJ: Prentice-Hall.

Kahneman, D. (2003). A perspective on judgement and choice: Mapping bounded rationality. *American Psychologist, 58*, pp. 697–720.

Kahneman, D., & Klein, G. (2009). Conditions for intuitive expertise: A failure to disagree. *American Psychologist, 64*, pp. 515–26.

Kahneman, D., & Tversky, A. (1979). Prospect theory: An analysis of decision under risk. *Econometrica: Journal of the Econometric Society, 47*, pp. 263–91.

Kamleitner, B., & Kirchler, E. (2007). Consumer credit use: A process model and literature review. *European Review of Applied Psychology, 57*, pp. 267–83.

Kaye, R., & Crowley, J. (2000). *Medical Device Use-Safety: Incorporating Human Factors Engineering into Risk Management*. Washington, DC: Food and Drug Administration.

Kevat, D.A., Cameron, P.A., Davies, A.R., Landrigan, C.P., & Rajaratnam, S.W. (2014). Safer hours for doctors and improved safety for patients. *The Medical Journal of Australia, 200*(7), pp. 396–8.

Kieras, D.E., & Bovair, S. (1984). The role of a mental model in learning to operate a device. *Cognitive Science, 8*, pp. 255–73.

Kim, S., Park, J., Han, S., & Kim, H. (2010). Development of extended speech act coding scheme to observe communication characteristics of human operators of nuclear power plans under abnormal conditions. *Journal of Loss Prevention in the Process Industries, 23*, pp. 539–48.

Kimmel, M. (2008). *Guyland: The Perilous World Where Boys Become Men*. New York. Harper.

Kirschenbaum, S.S. (1992). Influence of experience on information-search strategies. *Journal of Applied Psychology, 77*, pp. 343–52.

Klayman, J. (1988). Cue discovery in probabilistic environments: Uncertainty and experimentation. *Journal of Experimental Psychology: Learning, Memory, and Cognition, 14*, pp. 317–30.

Klein, G. (1989). Recognition-primed decisions. In W.B. Rouse (ed.), *Advances in Man-machine Systems Research* (vol. 5, pp. 47–92). Greenwich, CT: JAI Press.

Klein, G.A. (1993). A recognition primed decision (RPD) model of rapid decision making. In G.A. Kline, J. Orasanu, R. Caulderwood, & C.E. Zsambok (eds), *Decision Making in Action: Models and Methods* (pp. 138–47). Norwood, NJ: Ablex.

Klein, G. (1997). The recognition-primed decision (RPD) model: Looking back, looking forward. In C.E. Zsambok & G. Klein (eds), *Naturalistic Decision Making* (pp. 285–92). Mahwah, NJ: Lawrence Erlbaum Associates.

Klein, G. (1998). *Sources of Power: How People Make Decisions*. Cambridge, MA: MIT Press.

Klein, G. (2008). Naturalistic decision making. *Human Factors, 50*, pp. 456–60.

Klein, G., & Baxter, H.C. (2009). Cognitive transformation theory: Contrasting cognitive and behavioral learning. In D. Schmorrow, J. Cohn, & D. Nicholson (eds), *The PSI Handbook of Virtual Environments for Training and Education* (pp. 50–65). Westport, CT: Praeger Security International.

Klein, G.A., & Hoffman, R.R. (1993). Seeing the invisible: Perceptual-cognitive aspects of expertise. In M. Rabinowitz (ed.), *Cognitive science foundations of instruction* (pp. 203–27). Hillsdale, NJ: Lawrence Erlbaum.

Klein, G., Calderwood, R., & Clinton-Cirocco, A. (2010). Rapid decision making on the fire ground: The original study plus a postscript. *Journal of Cognitive Engineering and Decision Making, 4*(3), pp. 186–209.

Klein, G., Snowden, D., & Pin, C. L. (2011). Anticipatory thinking. In K.L. Mosier & U.M. Fischer (eds), *Informed by knowledge* (pp. 235–46). New York: Psychology Press.

Kleiner, B.M., & Drury, C.G. (1998). The use of verbal protocols to understand and design skills-based tasks. *Human Factors and Ergonomics in Manufacturing, 8*, pp. 23–9.

Kleinmuntz, B. (1990). Why we still use our heads instead of formulas: Toward an integrative approach. *Psychological Bulletin, 107*, pp. 296–310.

Kohl, N., Larsen, A., Larsen, J., Ross, A., & Tiourine, S. (2007). Airline disruption management – Perspectives, experiences and outlook. *Journal of Air Transport Management, 13*, pp. 149–62.

Kohn, L.T., Corrigan, J.M., & Donaldson, M.S. (eds). (2000). *To Err is Human: Building a Safer Health System* (vol. 627). Washington, DC: National Academies Press.

Konecni, V.J., & Ebbesen, E.B. (1982). An analysis of sentencing. In V.J. Konecni & E.B. Ebbesen (eds), *The Criminal Justice System: A Social-psychological Analysis* (pp. 189–229). San Francisco, CA: Freeman.

Konkola, R., Toumi-Grohn, T., Lambert, P., & Ludvigsen, S. (2007). Promoting learning and transfer between school and work. *Journal of Education and Work, 20*, pp. 211–28.

Kontogiannis, T., & Linou, N. (2001). Making instructions "visible" on the interface: An approach to learning fault diagnosis skills through guided discovery. *International Journal of Human-Computer Studies, 54*, pp. 53–79.

Krahmer, E., & Swerts, M. (2005). How children and adults produce and perceive uncertainty in audiovisual speech. *Language and Speech, 48*, pp. 29–54.

Kraut, R.E., Fussell, S.R., & Siegel, J. (2003). Visual information as a conversational resource in collaborative physical tasks. *Human-Computer Interaction, 18*, pp. 13–49.

Kuipers, A., Kappers, A., van Holten, C.R., van Bergen, J.H.W., & Oosterveld, W.J. (1990). *Spatial Disorientation Incidents in the R.N.L.A.F. F16 and F5 Aircraft and Suggestions for Prevention Awareness in Aerospace Operations* (AGARD-CP-478) (pp. Ov/E/1–Ov/E/16). Neuilly Sur Seine: NATO – AGARD.

Ladd, D.R. (1996). *Intonational Phonology*. Cambridge: Cambridge University Press.

Lan, S., Clarke, J.P., & Barnhart, C. (2006). Planning for robust airline operations: Optimizing aircraft routings and flight departure times to minimize passenger disruptions. *Transportation Science*, *40*, pp. 15–28.

Lavie, T., & Meyer, J. (2010). Benefits and costs of adaptive user interfaces. *International Journal of Human-Computer Studies*, *68*, pp. 508–24.

Lawrence, B.S. (2006). Organizational reference groups: A missing perspective on social context. *Organization Science*, *17*, pp. 80–100.

Laxmisan, A., Hakimzada, F., Sayan, O.R., Green, R.A., Zhang, J., & Patel, V.L. (2007). The multitasking clinician: Decision-making and cognitive demand during and after team handoffs in emergency care. *International Journal of Medical Informatics*, *76*, pp. 801–11.

Lazarus, R.S., & Folkman, S. (1984). *Stress, Appraisal, and Coping*. New York: Springer-Verlag.

Leake, D.B. (1999). Case-based reasoning. In W. Betchell & G. Graham (eds), *A Companion to Cognitive Science* (pp. 465–76). Malden, MA: Blackwell.

Lederer, P.J., & Nambimadom, R.S. (1998). Airline network design. *Operations Research*, *46*, pp. 785–804.

Lee, Y.-S., & Vakoch, D.A. (1996). Transfer and retention of implicit and explicit learning. *British Journal of Psychology*, *87*, pp. 637–51.

Lehto, M.R., & Nah, F. (2006). Decision-making models and decision support. In G. Salvendy (ed.), *Handbook of Human Factors and Ergonomics* (pp. 191–242). New York: John Wiley and Sons.

Lewis, A., Hall, T.E., & Black, A. (2011). Career stages in wildland firefighting: Implications for voice in risky situations. *International Journal of Wildland Fire*, *20*, pp. 115–24.

Lippa, K.D., Klein, H.A., & Shalin, V.L. (2008). Everyday expertise: Cognitive demands in diabetes self-management. *Human Factors*, *50*, pp. 112–20.

Lipshitz, R., Omodei, M., McLennan, J., & Wearing, A. (2007). What's burning? The RAWFS heuristic on the fire ground. In R. Hoffman (ed.), *Expertise Out of Context* (pp. 97–112). Mahwah, NJ: Lawrence Erlbaum.

Logan, G.D., & Schneider, D.W. (2006). Interpreting instructional cues in task switching procedures: The role of mediator retrieval. *Journal of Experimental Psychology: Learning, Memory, and Cognition*, *32*, 347–63.

Lois, J. (2001). Peaks and valleys: The gendered emotional culture of edgework. *Gender and Society*, *15*, pp. 381–406.

Louis, M.R. (1986). An investigator's guide to workplace culture. *Perspectives*, *3*, pp. 73–93.

Loveday, T., Wiggins, M.W., & Searle, B.J. (2012). Cue utilization and broad indicators of workplace expertise. *Journal of Cognitive Engineering and Decision-Making*, *8*, pp. 98–113.

Loveday, T., Wiggins, M.W., Harris, J., Smith, N., & O'Hare, D. (2013). An objective approach to identifying diagnostic experitise amongst power system controllers. *Human Factors, 55*, pp. 90–107.

Loveday, T., Wiggins, M.W., Searle, B.J., Festa, M., & Schell, D. (2013). The capability of static and dynamic features to distinguish competent from genuinely expert practitioners in pediatric diagnosis. *Human Factors, 55*, pp. 125–37.

Lowe, R.K. (2001). Components of expertise in the perception and interpretation of meteorological charts. In R.R. Hoffman & A.B. Markman (eds), *Interpreting Remote Sensory Imagery: Human Factors* (pp. 185–206). Boca Raton, FL: CRC Press LLC.

Lowe, R., & Boucheix, J.M. (2011). Cueing complex animations: Does direction of attention foster learning processes? *Learning and Instruction, 21*, 650–63.

Ma, R., & Kaber, D.B. (2007). Situational awareness and driving performance in a simulated navigation task. *Ergonomics, 50*, pp. 1351–64.

Mackworth, N.H. (1948). The breakdown of vigilance during prolonged visual search. *Quarterly Journal of Experimental Psychology, 1*, pp. 6–21.

Marcus, L.J., Dorn, B.C., & Henderson, J.M. (2006). Meta-leadership and national emergency preparedness: A model to build government connectivity. *Biosecurity and Bioterrorism, 4*, pp. 128–34.

Marino, C.J., & Mahan, R.P. (2005). Configural displays can improve nutrition-related decisions: An application of the proximity compatibility principle. *Human Factors, 47*, pp. 121–30.

Marshall, S. (2013). The use of cognitive aids during emergencies in anesthesia: A review of the literature. *Anesthesia & Analgesia, 117*, pp. 1162–71.

Matthews, M.D., Strater, L.D., & Endsley, M.R. (2004). Situational awareness requirements for infantry platoon leaders. *Military Psychology, 16*, pp. 149–61.

McCormack, C., Wiggins, M.W., Loveday, T., & Festa, M. (2014). Expert and competent non-expert visual cues during simulated diagnosis in intensive care. *Frontiers in Cognition, 5*, p. 949.

McInnes, F., & Attwater, D. (2004). Turn-taking and grounding in spoken telephone number transfers. *Speech Communication, 43*, pp. 205–23.

McNamara, G., & Bromiley, P. (1999). Risk and return in organizational decision making. *Academy of Management Journal, 42*, pp. 330–39.

McNamara, G., Moon, H., & Bromiley, P. (2002). Banking on commitment: Intended and unintended consequences of an organization's attempt to attenuate escalation of commitment. *Academy of Management Journal, 45*, pp. 443–52.

Meehl, P. (1954). *Clinical Versus Statistical Prediction: A Theoretical Analysis and a Review of the Evidence*. Minneapolis, MN: University of Minnesota Press.

Mendoza-Denton, R., Downey, G., Purdie, V. J., Davis, A., & Pietrzak, J. (2002). Sensitivity to status-based rejection: Implications for African American students' college experience. *Journal of Personality and Social Psychology, 83*, pp. 896–918.

Miller, G. A. (1956). The magical number seven, plus or minus two: Some limits on our capacity for processing information. *Psychological Review*, *101*, pp. 343–52.

Mitka, M. (2013). Joint commission warns of alarm fatigue: Multitude of alarms from monitoring devices problematic. *Journal of the American Medical Association*, *309*, pp. 2315–16.

Molesworth, B., Wiggins, M. W., & O'Hare, D. (2006). Improving pilots' risk assessment skills in low-flying operations: The role of feedback and experience. *Accident Analysis & Prevention*, *38*, pp. 954–60.

Montgomerie, J. (2006). Giving credit where it's due: Public policy and household debt in the United States, the United Kingdom and Canada. *Policy and Society*, *25*, pp. 109–41.

Morrison, B. W., Wiggins, M. W., & Porter, G. (2010). User preference for a control-based reduced processing decision support interface. *International Journal of Human Computer Interaction*, *26*, pp. 297–316.

Morrison, B. W., Morrison, N. M. V., Morton, J., & Harris, J. (2013). Using critical-cue inventories to advance virtual patient technologies in psychological assessment. Paper presented at the *Proceedings of the 25th Australian Computer-Human Interaction Conference*: Augmentation, Application, Innovation, Collaboration (OzChi'13).

Morrison, B. W., Wiggins, M. W., Bond, N., & Tyler, M. (2013). Measuring cue strength as a means of identifying an inventory of expert offender profiling cues. *Journal of Cognitive Engineering and Decision Making*, *7*, pp. 211–26.

Morrow, D., Rodvold, M., & Lee, A. (1994). Nonroutine transaction in controller-pilot communication. *Discourse Processes*, *17*, pp. 235–58.

Morrow, D. G., Miller, L. M. S., Ridolfo, H. E., Magnor, C., Fischer, U. M., Kokayeff, N. K., et al. (2009). Expertise and age differences in pilot decision making. *Aging, Neuropsychology, and Cognition*, *16*, pp. 33–55.

Mosier, K.L., Sethi, N., McCauley, S., Khoo, L., & Orasanu, J. M. (2007). What you don't know can hurt you: Factors impacting diagnosis in the automated cockpit. *Human Factors*, *49*, pp. 300–310.

Mukai, I., Kandy, B., Kesavabhotla, K., & Ungerleider, L. G. (2011). Exogenous and endogenous attention during perceptual learning differentially affect post-training target thresholds. *Journal of Vision*, *11*, pp. 1–15.

Mukai, I., Kim, D., Fukunaga, M., Japee, S., Marrett, S., & Ungerleider, L.G. (2007). Activations in visual and attention-related areas predict and correlate with the degree of perceptual learning. *Journal of Neuroscience*, *27*, pp. 11401–11.

Müller, S., Abernethy, B., & Farrow, D. (2006). How do world-class cricket batsmen anticipate a bowler's intention? *The Quarterly Journal of Experimental Psychology*, *59*, pp. 2162–86.

Munro, R. (1999). The cultural performance of control. *Organization Studies*, *20*, pp. 619–40.

Murphy, P., & Hall, K. (2008). *Learning and Practice: Agency and Identities*. London: Sage.

National Centre for Missing & Exploited Children. (2011). *Annual report.* Alexandria, VA: National Centre for Missing & Exploited Children.

Nembhard, I. M., & Edmondson, A.C. (2006). Making it safe: The effects of leader inclusiveness and professional status on psychological safety and improvement efforts in health care teams. *Journal of Organizational Behaviour, 27,* pp. 941–66.

Nijskens, R., & Wagner, W. (2011). Credit risk transfer activities and systemic risk: How banks became less risky individually but posed greater risks to the financial system at the same time. *Journal of Banking & Finance, 35,* pp. 1391–98.

Nisbett, R.E., & Wilson, T.D. (1977). Telling more than we can know: Verbal reports on mental processes. *Psychological Review, 84,* pp. 231–59.

Noakes, T.D., Gibson, A.S.C., & Lambert, E.V. (2005). From catastrophe to complexity: A novel model of integrative central neural regulation of effort and fatigue during exercise in humans: Summary and conclusions. *British Journal of Sports Medicine, 39,* pp. 120–24.

Norman, G., Young, M., & Brooks, L. (2007). Non-analytical models of clinical reasoning: The role of experience. *Medical Education, 41*(12), pp. 1140–45.

Norman, G.R., Coblentz, G.L., Brooks, L.R., & Babcook, C. J. (1992). Expertise in visual diagnosis: A review of the literature. *Academic Medicine, 67*(10 Suppl), S78–83.

Nosrati, H., Clay-Williams, R., Cunningham, F., Hillman, K., & Braithwaite, J. (2013). The role of organisational and cultural factors in the implementation of system-wide interventions in acute hospitals to improve patient outcomes: Protocol for a systematic literature review. *British Medical Journal Open, 3*(3).

Nygaard, L.C., & Queen, J.S. (2008). Communicating emotion: Linking affective prosody and word meaning. *Journal of Experimental Psychology: Human Perception and Performance, 34,* pp. 1017–30.

Nygaard, L.C., Herold, D.S., & Namy, L.L. (2009). The semantics of prosody: Acoustic and perceptual evidence of prosodic correlates to word meaning. *Cognitive Science, 33,* pp. 127–46.

O'Brien, K.S. & O'Hare, D. (2007). Situational awareness ability and cognitive skills training in a complex real-world task. *Ergonomics, 50,* pp. 1064–91.

O'Hare, D. (1997). Cognitive ability determinants of elite pilot performance. *Human Factors, 39,* pp. 540–52.

O'Hare, D., Mullen, N., Wiggins, M., & Molesworth, B. (2008). Finding the right case: The role of predictive features in memory for aviation accidents. *Applied Cognitive Psychology, 22,* pp. 1163–80.

O'Hare, D., Wiggins, M., Batt, R., & Morrison, D. (1994). Cognitive failure analysis for aircraft accident investigation. *Ergonomics, 37,* pp. 1855–69.

O'Hare, D., Wiggins, M.W., Williams, M., & Wong, W. (1998). Cognitive task analyses for decision centred design and training. *Ergonomics, 41,* pp. 1698–718.

O'Reilly, C. A. (1983). The use of information in organizational decision making: A model and some propositions. *Research in Organizational Behavior*, *5*, pp. 103–39.

Omodei, M.M., & Wearing, A.J. (1994). Perceived difficulty and motivated cognitive effort in a computer-simulated forest firefighting task. *Perceptual and Motor Skills*, *79*, pp. 115–27.

Ong, M., Russell, P.N., & Helton, W.S. (2013). Frontal cerebral oxygen response as an indicator of initial attention effort during perceptual learning. *Experimental Brain Research*, *229*, pp. 571–8.

Orasanu, J. (1997). Stress and naturalistic decision making: Strengthening the weak links. In R. Flin (ed.), *Decision making stress under stress: Emerging themes and applications* (pp. 49–160). Aldershot: Ashgate.

Orasanu, J. (2005). Crew collaboration in space: A naturalistic decision making perspective. *Aviation, Space, and Environmental Medicine*, *76*, B154–63.

Orasanu, J., & Connolly, T. (1993). The reinvention of decision making. In G.A. Klein, J. Orasanu, R. Calderwood, & C.E. Zsambok (eds), *Decision Making in Action: Models and Methods* (pp. 3–20). Norwood, NJ: Ablex.

Owen, C. (2001). The role of organisational context in mediating workplace learning and performance. *Computers in Human Behaviour*, *17*, pp. 597–614.

Owen, C. (2005). Beyond teamwork! Reconceptualising communication, co-ordination and collaboration in Air Traffic Control. *Human Factors and Aerospace Safety: An International Journal*, *4*, pp. 289–306.

Owen, C. (2008). Analysing joint work between activity systems. *Activités*, *5*(2), pp. 52–69.

Owen, C. (2009a). Instructor beliefs and their mediation of instructor strategies. *Journal of Workplace Learning*, *21*, pp. 477–95.

Owen, C. (2009b). Near-misses and mistakes in risky work: An exploration of work practices in high-3 environments. In C. Owen, P. Beguin, & G. Wackers (eds), *Risky Work Environments: Reappraising Human Work within Fallible Systems* (pp. 262–97). Aldershot: Ashgate.

Owen, C. (2013). Gendered communication and public safety: Women, men and incident management. *Australian Journal of Emergency Management*, *28*, 4–15.

Owen, C. (ed.). (2014a). *Enhancing Individual and Team Performance in Fire and Emergency Services*. Aldershot: Ashgate.

Owen, C. (2014b) *Ghosts in the Machine: Organisational Culture and Air Traffic Control*. Aldershot: Ashgate.

Owen, C., & Page, W. (2010). The reciprocal development of expertise in air traffic control. *International Journal of Applied Aviation Studies*, *10*, pp. 131–52.

Paas, F., Renkl, A., & Sweller, J. (2004). Cognitive load theory: Instructional implications of the interaction between information structures and cognitive architecture. *Instructional Science*, *32*, pp. 1–8.

Paas, F., Tuovinen, J.E., Tabbers, H., & Van-Gerven, P.W.M. (2003). Cognitive load measurement as a means to advance cognitive load theory. *Educational Psychologist, 38*, pp. 63–71.

Pachur, T., & Marinello, G. (2013). Expert intuitions: How to model the decision strategies of airport customs officers? *Acta Psychologica, 144*, pp. 97–103.

Parasuraman, R. (1984). The psychobiology of sustained attention. In J.S. Warm (ed.), *Sustained Attention in Human Performance* (pp. 61–101). Chichester: John Wiley & Sons.

Parasuraman, R., & Wickens, C.D. (2008). Humans: Still vital after all these years of automation. *Human Factors, 50*, pp. 511–20.

Park, J., & Jung, W. (2004). A study on the systematic framework to develop effective diagnosis procedures for nuclear power plants. *Reliability Engineering and System Safety, 84*, pp. 319–35.

Patel, V.L., & Groen, G.J. (1991). The general and specific nature of medical expertise: A critical look. In A. Ericsson and J. Smith (eds), *Towards a General Theory of Expertise: Prospects and Limits* (pp. 93–125). Cambridge: Cambridge University Press.

Patrick, J., Grainger, L., Gregov, A., Halliday, P., Handley, J., James, N., & O'Reilly, S. (1999). Training to break the barriers of habit in reasoning about unusual faults. *Journal of Experimental Psychology: Applied, 5*, pp. 314–35.

Patterson, R.E., Pierce, B.J., Bell, H.H., & Klein, G. (2010). Implicit learning, tacit knowledge, expertise development, and naturalistic decision making. *Journal of Cognitive Engineering and Decision Making, 4*, pp. 289–303.

Paul, C.L., & Sanson-Fisher, R.W. (1996). Experts' agreement on the relative effectiveness of 29 smoking reduction strategies. *Preventative Medicine, 25*, pp. 517–26.

Payne, R. (1991). Taking stock of corporate culture. *Personnel Management, 23*, pp. 26–30.

Pearson, R.W., Ross, M., & Dawes, R.M. (1992). Personal recall and the limits of retrospective questions in surveys. In J. Tanur (ed.), *Questions about survey questions* (pp. 65–94). Beverly Hills, CA: Sage.

Perry, N.C., Stevens, C.J., Wiggins, M.W., & Howell, C.E. (2007). Cough once for danger: Icons versus abstract warnings as informative alerts in civil aviation. *Human Factors, 49*, pp. 1061–71.

Perry, N.C., Wiggins, M.W., Childs, M., & Fogarty, G. (2012). Can reduced processing decision support interfaces improve the decision-making of less-experienced incident commanders? *Decision Support Systems, 52*, pp. 497–504.

Perry, N.C., Wiggins, M.W., Childs, M., & Fogarty, G. (2013). The application of reduced-processing decision support systems to facilitate the acquisition of decision-making skills. *Human Factors, 55*, pp. 535–44.

Petersen, M.A., & Rajan, R.G. (1995). The effect of credit market competition on lending relationships. *The Quarterly Journal of Economics, 110*, pp. 407–43.

Piske, T., Flege, J.E., MacKay, I.R.A., & Meador, D. (2002). The production of English vowels by fluent early and late Italian-English bilinguals. *Phonetica*, *59*, pp. 49–71.

Plessner, H., & Haar, T. (2006). Sports performance judgements from a social cognitive perspective. *Psychology of Sport and Exercise*, *7*, pp. 555–75.

Plessner, H., Schweizer, G., Brand, R., & O'Hare, D. (2009). A multiple-cue learning approach as the basis for understanding and improving soccer referees decision making. *Progress in Brain Research*, *174*, pp. 151–8.

Podsakoff, P.M., MacKenzie, S.B., Lee, J.Y., & Podsakoff, N.P. (2003). Common method biases in behavioral research: A critical review of the literature and recommended remedies. *Journal of Applied Psychology*, *88*, pp. 879–903.

Powell, W.W. (1990). The transformation of organizational forms: How useful is organization theory in accounting for social change. In R. Friedland & A.F. Robertson (eds), *Beyond the Marketplace: Rethinking Economy and Society* (pp. 301–29). New York: Aldine de Gruyter.

Prinzo, O.V. (1996). *An Analysis of Approach Control/Pilot Voice Communications* (Report No. DOT/FAA/AM-96/29). Washington, DC: Federal Aviation Administration.

Prinzo, O.V. (2008). The computation and effects of air traffic control message complexity and message length on pilot readback performance. *Proceedings of Measuring Behavior 2008, Maastricht, The Netherlands*, pp. 188–9.

Prinzo, O.V., & Britton, T.W. (1993). *ATC/Pilot Voice Communications: A Survey of the Literature* (No. DOT/FAA/AM-93/20). Washington, DC: Office of Aviation Medicine.

Prowse, G., Palmisano, S., & Favelle, S. (2008). Time-to-contact perception during simulated night landing. *The International Journal of Aviation Psychology*, *18*, pp. 207–23.

Ramnarayan, P., Tomlinson, A., Rao, A., Coren, M., Winrow, A., & Britto, J. (2003). ISABEL: A web-based differential diagnostic aid for paediatrics: Results from an initial performance evaluation. *Archives of Disease in Childhood*, *88*(5), pp. 408–13.

Ranyard, R., Hinkley, L., Williamson, J., & McHugh, S. (2006). The role of mental accounting in consumer credit decision processes. *Journal of Economic Psychology*, *27*, pp. 571–88.

Rasmussen, J. (1983). Skills, rules, and knowledge: Signals, signs, and symbols, and other distinctions in human performance models. *IEEE Transactions on Systems, Man and Cybernetics*, *13*, pp. 257–66.

Rasmussen, J. (1986). *Information Processing and Human-machine Interaction: An Approach to Cognitive Engineering*. New York: Elsevier.

Rasmussen, J. (1993). Diagnostic reasoning in action. *IEEE Transactions on Systems, Man and Cybernetics*, *23*, pp. 981–92.

Reason, J. (2000). Human error: Models and management. *British Medical Journal*, *320*, pp. 768–70.

Rehder, B. (2001). Interference between cognitive skills. *Journal of Experimental Psychology: Learning, Memory, and Cognition, 27*, pp. 451–69.

Reischman, R. R., & Yarandi, H. N. (2002). Critical care cardiovascular nurse expert and novice diagnostic cue utilization. *Journal of Advanced Nursing, 39*, pp. 24–34.

Reuber, R. (1997). Management experience and management expertise. *Decision Support Systems, 21*, pp. 51–60.

Richards, D., & Compton, P. (1998). Taking up the situated cognition challenge with ripple down rules. *International Journal of Human-Computer Studies, 49*, pp. 895–926.

Ridgway, S., Carder, D., Finneran, J., Keogh, M., Kamolnick, T., Todd, M., & Goldblatt, A. (2006). Dolphin continuous auditory vigilance for five days. *Journal of Experimental Biology, 209*, pp. 3621–8.

Ridgway, S., Keogh, M., Carder, D., Finneran, J., Kamolnick, T., Todd, M., & Goldblatt, A. (2009). Dolphins maintain cognitive performance during 72 to 120 hours of continuous auditory vigilance. *Journal of Experimental Biology, 212*, pp. 1519–27.

Rieh, S. Y. (2002). Judgement of information quality and cognitive authority in the web. *Journal of the American Society for Information Science and Technology, 53*, pp. 145–61.

Rodgers, W. (1999). The influences of conflicting information on novices and loan officers' actions. *Journal of Economic Psychology, 20*, pp. 123–45.

Rokeach, M.J. (1973). *The Nature of Human Values*. New York: Free Press.

Roscoe, S.N., Corl, L., & LaRoche, J. (2001). *Predicting Human Performance* (5th ed.). Pierrefonds, CA: Helio Press.

Ross, H.A., Russell, P.N., & Helton, W.S. (2014). Effects of breaks and goal switches on the vigilance decrement. *Experimental Brain Research, 232*, pp. 1729–37.

Rouse, W.B., & Morris, N.M. (1986). On looking into the black box: Prospects and limits in the search for mental models. *Psychological Bulletin, 100*, pp. 349–63.

Rowe, R., Horswill, M.S., Kronvall-Parkinson, M., Poulter, D.R., & Mckenna, F.P. (2009). The effect of disguise on novice and expert tennis players anticipation ability. *Journal of Applied Sport Psychology, 21*, pp. 178–85.

Rowe, R.M., & McKenna, F.P. (2001). Skilled acquisition in real-world tasks: Measurement of attentional demands in the domain of tennis. *Journal of Experimental Psychology: Applied, 7*, pp. 60–67.

Ruchala, L.V., Hill, J.W., & Dalton, D. (1996). Escalation and the diffusion of responsibility: A commercial lending experiment. *Journal of Business Research, 37*, pp. 15–26.

Saariluoma, P. (1985). Chess players' intake of task-relevant cues. *Memory & Cognition, 13*, pp. 385–91.

Saariluoma, P. (1990). Apperception and restructuring in chess players' problem solving. In K.J. Gilhooly, M.T.G. Keane, R.H. Logie, & G. Erdos (eds),

Lines of Thought: Reflections on the Psychology of Thinking (pp. 41–57). London: Wiley.

Salas, E., Rosen, M.A., Burke, C.S., Goodwin, G.F., & Fiore, S.M. (2006). The making of a dream team: When expert teams do best. In K.A. Ericsson, N. Charness, P.J. Feltovich, & R.R. Hoffman (eds), *The Cambridge Handbook of Expertise and Expert Performance* (pp. 439–53). New York: Cambridge University Press.

Salfati, C.G., & Canter, D.V. (1999). Differentiating stranger murders: Profiling offender characteristics from behavioral styles. *Behavioral Sciences and the Law, 17*, pp. 391–406.

Salmon, P.M., Stanton, N.A., Walker, G.H., Jenkins, D., Ladva, D., Rafferty, L., & Young, M. (2009). Measuring situational awareness in complex systems: Comparison of measures study. *International Journal of Industrial Ergonomics, 39*, pp. 450–500.

Salzwedel, C., Bartz, H. J., Kühnelt, I., Appel, D., Haupt, O., Maisch, S., & Schmidt, G. N. (2013). The effect of a checklist on the quality of post-anaesthesia patient handover: A randomized controlled trial. *International Journal for Quality in Health Care, 25*, pp. 176–81.

Santilla, P., Korpela, S., & Häkkanen, H. (2004). Expertise and decision-making in the linking of car crime series. *Psychology, Crime and Law, 10*, pp. 97–112.

Sarasvathy, D.K., Simon, H.A., & Lave, L. (1998). Perceiving and managing business risks: Differences between entrepreneurs and bankers. *Journal of Economic Behavior & Organization, 33*, pp. 207–25.

Sarter, N.B., & Woods, D.D. (1991). Situational awareness: A critical but ill-defined phenomenon. *The International Journal of Aviation Psychology, 1*, pp. 45–57.

Sarter, N., & Woods, D.D. (1995). How in the world did I ever get into that mode? Mode error and awareness in supervisory control. *Human Factors, 37*, pp. 5–19.

Sauer, J., Wastell, D. G., Hockey, G.R.J., Crawshaw, C.M., Ishak, M., & Downing, J.C. (2002). Effects of display design on performance in a simulated ship navigation environment. *Ergonomics, 45*, pp. 329–47.

Schacter, D. (1989). Memory. In M. Posner (ed.), *Foundations of Cognitive Science* (pp. 638–725). Cambridge, MA: MIT Press.

Scheeres, H., & Rhodes, C. (2006). Between cultures: Values, training and identity in a manufacturing firm. *Journal of Organizational Change Management, 19*, pp. 223–37.

Schein, E.H. (1996). Culture: The missing concept in organization studies. *Administrative Science Quarterly, 41*, pp. 229–40.

Schleburg, M., Christiansen, L., Thornhill, N.F., & Fay, A. (2013). A combined analysis of plant connectivity and alarm logs to reduce the number of alerts in an automation system. *Journal of Process Control, 23*, pp. 839–51.

Schmidt, H.G., & Boshuizen, H.P. (1993). On acquiring expertise in medicine. *Educational Psychology Review, 5*, pp. 205–21.

Schneider, T.R. (2004). The role of neuroticism on psychological and physiological stress responses. *Journal of Experimental Social Psychology*, *40*, pp. 795–804.

Schneider, T.R. (2008). Evaluations of stressful transactions: What's in an appraisal? *Stress and Health*, *24*, pp. 151–8.

Schneider, T.R., Rench, T.A., Lyons, J.B., & Riffle, R.R. (2012). The influence of neuroticism, extraversion, and openness on stress responses. *Stress and Health*, *28*, pp. 102–10.

Schneider, V.I., Healy, A.F., & Barshi, I. (2004). Effects of instruction modality and readback on accuracy in following navigation commands. *Journal of Experimental Psychology*, *10*, pp. 245–57.

Schraagen, J.M., & Leijenhorst, H. (2001). Searching for evidence: Knowledge and search strategies used by forensice scientists. In E. Salas & G. A. Kline (eds), *Linking Expertise and Naturalistic Decision Making* (pp. 263–74). Mahwah, NJ: Lawrence Erlbaum.

Schriver, A.T., Morrow, D.G., Wickens, C.D., & Talleur, D.A. (2008). Expertise differences in attentional strategies related to pilot decision making. *Human Factors*, *50*, pp. 864–78.

Schumaker, R.M., & Czerwinski, M.P. (1990). Mental models and the acquisition of expert knowledge. In R.R. Hoffman (ed.), *The Psychology of Expertise* (pp. 61–79). New York: Springer-Verlag.

Schwarz, N. (1990). *Feelings as Information: Informational and Motivational Functions of Affective Atates*. New York: Guilford Press.

Schwarz, N., & Clore, G.L. (1988). How do I feel about it? The informative function of affective states. In K. Fielder & J.P. Forgas (eds), *Affect, Cognition, and Social Behaviour* (pp. 44–62). Toronto: Hogrefe.

Scott, I.A. (2009). Errors in clinical reasoning: causes and remedial strategies. *British Medical Journal*, *338*.

Sebanz, N., & Shiffrar, M. (2009). Detecting deception in a bluffing body: The role of expertise. *Psychonomic Bulletin and Review*, *16*, pp. 170–75.

See, J.E., Howe, S.R., Warm, J.S., & Dember, W.N. (1995). Meta-analysis of the sensitivity decrement in vigilance. *Psychological Bulletin*, *117*, pp. 230–49.

Sendra, V.C., Kaland, C., Swerts, M., & Prieto, P. (2013). Perceiving incredulity: The role of intonation and facial gestures. *Journal of Pragmatics*, *47*, pp. 1–13.

Shah, A.K., & Oppenheimer, D.M. (2008). Heuristics made easy: An effort-reduction framework. *Psychological Bulletin*, *134*, pp. 207–22.

Shanmugam, B., & Bourke, P. (1992). Biases in appraising creditworthiness. *International Journal of Bank Marketing*, *10*(3), pp. 10–16.

Shanteau, J. (1988). Psychological characteristics and strategies of expert decision-makers. *Acta Psychologica*, *68*, pp. 203–15.

Shanteau, J. (1991). Psychological characteristics and strategies of experts. In G. Wright & F. Bolger (eds), *Expertise and Decision Support* (pp. 55–76). New York: Plenum.

Shanteau, J. (1992). How much information does an expert use? Is it relevant? *Acta Psychologica*, *81*, pp. 75–86.

Shapiro, K.L., & Raymond, J.F. (1989). Training of efficient occularmotor strategies enhance skill aquisition. *Acta Psychologica, 17*, pp. 217–42.

Sheridan, T.B. (2008). Risk, human error, and system resilience: Fundamental ideas. *Human Factors, 50*, pp. 418–26.

Sherif, M. (1935). A study of some social factors in perception. *Archives of Psychology, 27*, #18.

Shiffrin, R.M., & Schneider, W. (1977). Controlled and automatic human information processing: II. Perceptual learning, automatic attending and a general theory. *Psychological Review, 84*, p. 127.

Shim, J., Carlton, L. G., Chow, J.W., & Chae, W.S. (2005). The use of anticipatory visual cues by highly skilled tennis players. *Journal of Motor Behavior, 37*, pp. 164–75.

Siassakos, D., Timmons, C., Hogg, F., Epee, M., Marshall, L., & Draycott, T. (2009). Evaluation of a strategy to improve undergraduate experience in obstetrics and gynaecology. *Medical Education, 43*, pp. 669–73.

Sidaras, S.K., Alexander, J.E.D., & Nygaard, L.C. (2009). Perceptual learning of systematic variation in Spanish-accented speech. *Journal of the Acoustical Society of America, 125*, pp. 3306–16.

Simnett, R. (1996). The effect of information selection, information processing, and task complexity on predictive accuracy of auditors. *Accounting, Organizations, and Society, 21*, 699–719.

Simon, H.A. (1957). *Models of Man: Social and Rational.* New York: Wiley.

Simon, H.A., & Chase, W.G. (1973). Skill in chess: Experiments with chess-playing tasks and computer simulation of skilled performance throw light on some human perceptual and memory processes. *American Scientist*, pp. 394–403.

Skantze, G. (2005). Exploring human error recovery strategies: Implications for spoken dialogue systems. *Speech Communication, 45*, pp. 325–41.

Slovic, P. (1969). Analyzing the expert judge: A descriptive study of stockbroker's decision processes. *Journal of Applied Psychology, 53*, pp. 255–63.

Slovic, P., Finucane, M.L., Peters, E., & MacGregor, D.G. (2007). The affect heuristic. *European Journal of Operational Research, 177*, pp. 1333–52.

Small, A.J., Wiggins, M.W., & Loveday, T. (2014). Cue-based processing capacity, cognitive load and the completion of simulated short-duration vigilance tasks in power transmission control. *Applied Cognitive Psychology, 28*, pp. 481–7.

Smeeton, N.J., Williams, A.M., Hodges, N.J., & Ward, P. (2005). The relative effectiveness of various instructional approaches in developing anticipation skill. *Journal of Experimental Psychology: Applied, 11*, pp. 98–110.

Smiricich, L. (1983). Concepts of culture and organizational analysis. *Administrative Science Quarterly, 28*, pp. 339–58.

Smith, A.P., & Broadbent, D.E. (1981). Noise and levels of processing. *Acta Psychologica, 47*, pp. 129–42.

Smith, K., & Hancock, P.A. (1995). Situational awareness is adaptive, externally directed consciousness. *Human Factors, 37*, pp. 137–48.

Smith, V.L., & Clark, H.H. (1993). On the course of answering questions. *Journal of Memory and Language*, *32*, pp. 25–38.

Smitka, M. (1991). *Governance by Trust*. New York: Columbia University Press.

Snyder, C.W., Vandromme, M.J., Tyra, S.L., Porterfield Jr, J.R., Clements, R.H., & Hawn, M.T. (2011). Effects of virtual reality simulator training method and observational learning on surgical performance. *World Journal of Surgery*, *35*, pp. 245–52.

Sohn, Y.W., & Doane, S.M. (2004). Memory processes of flight situation awareness: Interactive roles of working memory capacity, long-term working memory, and expertise. *Human Factors*, *46*, pp. 461–75.

Song, J.H., Skoe, E., Banai, K., & Kraus, N. (2012). Training to improve hearing speech in noise: Biological mechanisms. *Cerebral Cortex*, *22*, pp. 1180–190.

Srinivasan, R.J., & Massaro, D.W. (2003). Perceiving from the face and voice: Distinguishing statements from echoic questions in English. *Language and Speech*, *46*, pp. 1–22.

Stanovich, K.E. (2010). *Decision Making and Rationality in the Modern World*. Oxford: Oxford University Press.

Stanton, N.A., & Young, M.S. (2000). A proposed psychological model of driving automation. *Theoretical Issues in Ergonomic Science*, *1*, pp. 315–31.

Stanton, N.A., Salmon, P.M., Walker, G.H., Baber, C., & Jenkins, D.P. (2005). *Human Factors Methods: A Practical Guide for Engineering and Design*. Aldershot: Ashgate.

Stokes, A.F., Kemper, K., & Kite, K. (1997). Aeronautical decision making, cue recognition, and expertise under time pressure. In C.E. Zsambok & G. Klein (eds), *Naturalistic decision making* (pp. 183–96). Mahwah, NJ: Lawrence Erlbaum Associates.

Stokes, A.F., Kemper, K.L., & Marsh, R. (1992). *Time-stressed Flight Decision Making: A Study of Expert and Novice Aviators*. Savoy, IL: University of Illinois, Aviation Research Laboratory.

Stone, J., & Moskowitz, G.B. (2011). Non-conscious bias in medical decision making: What can be done to reduce it? *Medical Education*, *45*, pp. 768–76.

Sträter, O. (2003). Investigation of communication errors in nuclear power plants. In R. Dietrich (ed.), *Communication in High-risk Environments* (pp. 155–77). Hamburg: Helmut Buske Verlag.

Streitenberger, K., Breen-Reid, K., & Harris, C. (2006). Handoffs in care: Can we make them safer? *Pediatric Clinics of North America*, *53*, pp. 1185–95.

Sullivan, P., & Girginer, H. (2002). The use of discourse analysis to enhance ESP teacher knowledge: An example using aviation English. *English for Specific Purposes*, *21*, pp. 397–404.

Sutcliffe, K.M., & McNamara, G. (2001). Controlling decision-making practice in organizations. *Organization Science*, *12*, pp. 484–501.

Suvachittanont, W., Arnott, D.R., & O'Donnell, P.A. (1994). Adaptive design in executive information systems development: A manufacturing case study. *Journal of Decision Systems*, *3*, pp. 277–99.

Svenson, O. (2008). Decisions among time saving options: When intuition is strong and wrong. *Acta Psychologica, 127*, pp. 501–9.

Svensson, E.A.I., & Wilson, G.F. (2002). Psychological and psychophysiological models of pilot performance for systems development and mission evaluation. *The International Journal of Aviation Psychology, 12*, pp. 95–110.

Sweller, J. (1988). Cognitive load during problem solving: Effects on learning. *Cognitive Science, 12*, pp. 257–85.

Swerts, M., & Krahmer, E. (2005). Audiovisual prosody and feeling of knowing. *Journal of Memory and Language, 53*, pp. 81–94.

Taatgen, N.A., & Lee, F.J. (2003). Production compilation: A simple mechanism to model complex skill acquisition. *Human Factors, 45*, pp. 61–76.

Tajima, A. (2004). Fatal miscommunication: English in aviation safety. *World Englishes, 23*, pp. 451–70.

Thammasitboon, S., Thammasitboon, S., & Singhal, G. (2013). System-related factors contributing to diagnostic errors. *Current Problems in Pediatric and Adolescent Health Care, 43*, pp. 242–7.

Thompson, J.E., Collett, L.W., Langbart, M.J., Purcell, N.J., Boyd, S.M., Yuminaga, Y., Ossolinski, G., Susanto, C., & McCormack, A. (2011). Using the ISBAR handover tool in junior medical officer handover: A study in an Australian tertiary hospital. *Postgraduate Medical Journal, 87*, pp. 340–44.

Tibby, S.M., Correa-West, J., Durward, A., Ferguson, L., & Murdoch, I. A. (2004). Adverse events in a paediatric intensive care unit: Relationship to workload, skill mix and staff supervision. *Intensive Care Medicine, 30*, pp. 1160–66.

Todd, P.M., & Gigerenzer, G. (2003). Bounding rationality to the world. *Journal of Economic Psychology, 24*, pp. 143–65.

Tomaka, J., Blascovich, J., Kelsey, R.M., & Leitten, C.L. (1993). Subjective, physiological, and behavioral effects of threat and challenge appraisal. *Journal of Personality and Social Psychology, 65*(2), pp. 248–60.

Trice, H.M. & Beyer, J.M. (1984). Studying organization culture through rites and ceremonials. *Academy of Management Review, 9*, pp. 653–69.

Trude, A.M., & Brown-Schmidt, S. (2012). Talker-specific perceptual adaptation during online speech perception. *Language and Cognitive Processes, 27*, pp. 979–1001.

Trude, A.M., Tremblay, A., & Brown-Schmidt, S. (2013). Limitations on adaptation to foreign accents. *Journal of Memory and Language, 69*, pp. 349–67.

Tsang, P.S., & Vidulich, M. A. (2006). Mental workload and situation awareness. In G. Salvendy (eds), *Handbook of Human Factors and Ergonomics* (3rd ed., pp. 243–68). Hoboken, NJ: John Wiley & Sons.

Turvey, B.E. (2011). *Criminal Profiling: An Introduction to Behavioral Evidence Analysis*. Oxford: Academic Press.

Tversky, A., & Kahneman, D. (1973). Availability: A heuristic for judging frequency and probability. *Cognitive Psychology, 5*, pp. 207–32.

Tversky, A., & Kahneman, D. (1974). Judgement under uncertainty: Heuristics and biases. *Science, 185*, pp. 1124–31.

Tversky, A., & Kahneman, D. (1982). Evidential impact of base rates. In D. Kahneman, P. Slovic, & A. Tversky (eds), *Judgement Under Uncertainty: Heuristics and Biases* (pp. 153–60). Cambridge: Cambridge University Press.

Tversky, B., Morrison, J. B., & Betrancourt, M. (2002). Animation: Can it facilitate? *International Journal of Human Computer Studies, 57,* pp. 1–16.

Tyler, M., & Fairbrother, P. (2013). Bushfires are "men's business": The importance of gender and rural hegemonic masculinity. *Journal of Rural Studies, 30,* pp. 110–19.

Tzu, S., & Sawyer, R.D. (1994). *The Art of War.* New York: Basic Books.

Unkelbach, C., Forgas, J.P., & Denson, T.F. (2008). The turban effect: The influence of Muslim headgear and induced affect on aggressive responses in the shooter bias paradigm. *Journal of Experimental Social Psychology, 44,* pp. 1409–13.

Van Der Hulst, M., Meijman, T., & Rothengatter, T. (1999). Anticipation and the adaptive control of safety margins in driving. *Ergonomics, 42,* pp. 336–45.

Van Gog, T., Kester, L., & Paas, F. (2010). Effects of concurrent monitoring on cognitive load and performance as a function of task complexity. *Applied Cognitive Psychology, 25,* pp. 584–7.

Van Knippenberg, D. (2000). Work motivation and performance: A social identity perspective. *Applied Psychology: An International Review, 49,* pp. 357–71.

Verdonik, D. (2010). Between understanding and misunderstanding. *Journal of Pragmatics, 42,* pp. 1364–79.

Vicente, K.J., & Wang, J.H. (1998). An ecological theory of expertise effects in memory recall. *Psychological Review, 105,* pp. 33–57.

Victorian Bushfires Royal Commission (VBRC). (2010). *Final Report: Summary.* Melbourne: Parliament of Victoria.

Wadhera, R.K., Parker, S.H., Burkhart, H.M., Greason, K.L., Neal, J.R., Levenick, K.M., Wiehmann, D.A., & Sundt, T.M. (2010). Is the "sterile cockpit" concept applicable to cardiovascular surgery critical intervals or critical events? The impact of protocol-driven communication during cardiopulmonary bypass. *The Journal of Thoracic and Cardiovascular Surgery, 139,* pp. 312–19.

Warm, J.S., & Dember, W.N. (1998). Tests of the vigilance taxonomy. In R.R. Hoffman, M.F. Sherrick, & J.S. Warm (eds), *Viewing Psychology as a Whole: The Integrative Science of William N. Dember* (pp. 87–112). Washington, DC: American Psychological Association.

Watanabe, T., Náñez, J.E., & Sasaki, Y. (2001). Perceptual learning without perception. *Nature, 413,* pp. 844–8.

Watanabe, T., Nanez, J.E., Koyama, S., Mukai, I., Liederman, J., & Sasaki, Y. (2002). Greater plasticity in lower-level than higher-level visual motion processing in a passive perceptual learning task. *Nature Neuroscience, 5,* pp. 1003–9.

Weaver, S.J., Newman-Toker, D.E., & Rosen, M.A. (2012). Reducing cognitive skill decay and diagnostic error: Theory-based practices for continuing education in health care. *Journal of Continuing Education in the Health Professions, 32,* pp. 269–78.

Weick, K.E. (2001). *Making Sense of the Organization*. Oxford: Blackwell.

Weigand, E. (1999). Misunderstanding: The standard case. *Journal of Pragmatics*, *31*, pp. 763–85.

Weiss, D.J., & Shanteau, J. (2003). Empirical assessment of expertise: Quantitative formal models of human performance. *Human Factors*, *45*, pp. 104–16.

Westrum, R. (1993). Cultures with requisite imagination. In J.A.H. Wise, V.D. Hopkin, & P. Stager. (eds), *Verification and Validation of Complex Systems: Human Factors Issues* (pp. 401–16). Berlin: Springer-Verlag.

Wickens, C.D. (2008). Multiple resources and mental workload. *Human Factors*, *50*, pp. 449–55.

Wickens, C.D., & Carswell, C.M. (1995). The proximity compatibility principle: Its psychological foundation and relevance to display design. *Human Factors*, *37*, pp. 473–94.

Wickens, C.D., & Flach, J.M. (1988). Information processing. In E.L. Wiener & D.C. Nagel (eds), *Human Factors in Aviation* (pp. 111–56). San Diego, CA: Academic Press.

Wickens, C.D., & Hollands, J.G. (2000). *Engineering Psychology and Human Performance* (3rd ed.). Upper Saddle River, NJ: Prentice Hall.

Wickens, C.D., Hollands, J.G., Banbury, S., & Parasuraman, R. (2013). *Engineering Psychology and Human Performance* (4th ed.). Upper Saddle River, NJ: Pearson.

Wiggins, M.W. (2006). Cue-based processing and human performance. In W. Karwowski (ed.), *International Encyclopedia of Ergonomics and Human Factors* (2nd ed., pp. 641–5). London: Taylor & Francis.

Wiggins, M.W. (2012). The role of cue utilisation and adaptive interface design in the management of skilled performance in operations control. *Theoretical Issues in Ergonomics Science*, pp. 1–10.

Wiggins, M.W., & Bollwerk, S. (2006). Heuristic-based information acquisition and decision making amongst pilots. *Human Factors*, *48*, pp. 734–46.

Wiggins, M.W. & O'Hare, D. (1995). Expertise in aeronautical weather-related decision-making: A cross-sectional analysis of general aviation pilots. *Journal of Experimental Psychology: Applied*, *1*, pp. 305–20.

Wiggins, M.W., & O'Hare, D. (2003a). Weatherwise: Evaluation of a cue-based training approach for the recognition of deteriorating weather conditions during flight. *Human Factors*, *45*, pp. 337–45.

Wiggins, M.W., & O'Hare, D. (2003b). Expert and novice pilot perceptions of static in-flight images of weather. *International Journal of Aviation Psychology*, *13*, pp. 173–87.

Wiggins, M.W., & Stevens, C. (1999). *Aviation Social Science: Research Methods in Practice*. Aldershot: Ashgate.

Wiggins, M.W., Brouwers, S., Davies, J., & Loveday, T. (2014). Trait-based cue utilisation and initial skill aquisition: Implications for models of the progression to expertise. *Frontiers in Psychology*, *5*, pp. 1–8.

Wiggins, M.W., Harris, J.M., Loveday, T., & O'Hare, D. (2010). *Expert Intensive Skills Evaluation* (EXPERTise) 1.0 Software Package. Sydney: Macquarie University.

Wiggins, M.W., Stevens, C., Howard, A., Henley, I., & O'Hare, D. (2002). Expert, intermediate and novice performance during simulated pre-flight decision-making. *Australian Journal of Psychology*, *54*, pp. 162–67.

Wigton, R.S. (1996). Social judgement theory and medical judgement. *Thinking and Reasoning*, *2*, pp. 175–90.

Wilcox, M.E., Chong, C.A., Niven, D.J., Rubenfeld, G.D., Rowan, K.M., Wunsch, H., & Fan, E. (2013). Do intensivist staffing patterns influence hospital mortality following ICU admission? A systematic review and meta-analyses. *Critical Care Medicine*, *41*, pp. 2253–74.

Williams, A.M., Ward, P., Knowles, J.M., & Smeeton, N.J. (2002). Anticipation skill in a real-world task: Measurement, training, and transfer in tennis. *Journal of Experimental Psychology: Applied*, *8*, pp. 259–70.

Williams, C.A., Haslam, R.A., & Weiss, D.J. (2008). The Cochran-Weiss-Shanteau performance index as an indicator of upper limb risk assessment expertise. *Ergonomics*, *51*, pp. 1219–37.

Williams, J.E.D. (1967). *The Operation of Airliners*. London: Hutchinson and Co.

Williams, S.M. (1992). Putting case-based instruction into context: Examples from legal and medical education. *The Journal of the Learning Sciences*, *2*, pp. 367–427.

Wilson, K.M., Helton, W.S., & Wiggins, M.W. (2013). Cognitive engineering. WIRES: *Cognitive Science*, *4*, pp. 17–31.

Wiltshire, T.J., Neville, K.J., Lauth, M.R., Rinkinen, C., & Ramirez, L.F. (2014). Applications of cognitive transformation theory examining the role of sensemaking in the instruction of air traffic control students. *Journal of Cognitive Engineering and Decision Making*, *8*, pp. 219–47.

Witteman, M.J., Weber, A., & McQueen, J.M. (2013). Foreign accent strength and listener familiarity with an accent codetermine speed of perceptual adaptation. *Attention Perception & Psychophysics*, *75*, pp. 312–19.

Woods, D.D. (1988). Coping with complexity: The psychology of human behavior in complex systems. In L.P. Goldstein, H.B. Andersen, & S.E. Olsen (eds), *Tasks, Errors and Mental Models* (pp. 128–48). New York: Taylor & Francis.

Woods, D.D., & Roth, E.M. (1988). Cognitive engineering: Human problem-solving with tools. *Human Factors*, *30*, pp. 415–30.

Wouda, J.C., & van de Wiel, H.B.M. (2013). Education in patient-physician communication: How to improve effectiveness? *Patient Education and Counseling*, *90*, pp. 46–53.

Yeo, G.B., & Neal, A. (2004). A multilevel analysis of effort, practice, and performance: Effects of ability, conscientiousness, and goal orientation. *Journal of Applied Psychology*, *89*, pp. 231–47.

Yoon, S.O., & Brown-Schmidt, S. (2013). Lexical differentiation in language production and comprehension. *Journal of Memory and Language, 69,* pp. 397–416.

Zhang, Y., Drews, F.A., Westenskow, D.R., Foresti, S., Agutter, J., Bermudez, J.C., Blike, G., & Loeb, R. (2002). Effects of integrated graphical displays on situational awareness in anaesthesiology. *Cognition, Technology and Work, 4,* pp. 82–90.

Index